# COGENERATION MANAGEMENT REFERENCE GUIDE

F. William Payne, Editor

# COGENERATION
# MANAGEMENT
# REFERENCE GUIDE

F. William Payne, Editor

**Library of Congress Cataloging-in-Publication Data**

Payne, F. William, 1924-
    Cogeneration management reference guide / F. William Payne, editor.
    p. cm.
    Includes index.
    ISBN 0-88173-248-6
    1. Cogeneration of electric power and heat. I. Payne, F. William, 1924-
TK1071.C6327     1997     333.793--dc21     96-48537
                                             CIP

*Cogeneration Management Reference Guide.*

Published by The Fairmont Press, Inc.
700 Indian Trail
Lilburn, GA 30247

Printed in the United States of America

10 9 8 7 6 5 4 3 2 1

ISBN 0-88173-248-6 FP

ISBN 0-13-743261-5 PH

While every effort is made to provide dependable information, the publisher, authors, and
editors cannot be held responsible for any errors or omissions.

Distributed by Prentice Hall PTR
Prentice-Hall, Inc.
A Simon & Schuster Company
Upper Saddle River, NJ 07458

Prentice-Hall International (UK) Limited, London
Prentice-Hall of Australia Pty. Limited, Sydney
Prentice-Hall Canada Inc., Toronto
Prentice-Hall Hispanoamericana, S.A., Mexico
Prentice-Hall of India Private Limited, New Delhi
Prentice-Hall of Japan, Inc., Tokyo
Simon & Schuster Asia Pte. Ltd., Singapore
Editora Prentice-Hall do Brasil, Ltda., Rio de Janeiro

# Dedication

This book is dedicated to the creative—and tough—individuals who are responsible for the evolution of the cogeneration industry from a comparatively simple beginning to the powerful and complex activity which it now is.

Starting in the early 80s, I worked with cogeneration entrepreneurs and originators to report their activities in the quarterly journal *Strategic Planning for Energy and the Environment*, published under the auspices of The Association of Energy Engineers. Many of their early articles were then presented in book format in 1985.

Since then, cogeneration has vastly ramified. It was necessary, in 1986, to expand into a second publication, *The Cogeneration and Competitive Power Journal*, also sponsored by AEE, in order to report properly on this dynamic technology.

Events have proved that the roadblocks thrown up against the budding cogeneration industry, and against those individuals who doggedly persevered in developing it, were not only wrong, but futile.

The strengths and successes of cogeneration today can be traced directly to their pioneering efforts. In respectful appreciation, this book is dedicated to them.

# Table of Contents

# Acknowledgements

Each chapter in the *Cogeneration Management Reference Guide* is practical, original information, written by experienced authors with high qualifications in their fields.

Many of the reports were originally presented at past Cogeneration and Independent Power Congresses, or at World Energy Engineering Congresses, both sponsored by the Association of Energy Engineers. Others have appeared as original articles in *The Cogeneration and Competitive Power Journal*.

The openness of the authors of these chapters in sharing their experiences has helped advance the knowledge of cogeneration. We thank them for their support.

# List of Contributors

Richard D. Kennedy, P.E., C.E.M., C.C.P.
Cecelio E. Gracias
Lloyd S. Hoffsttater
James R. Geers
Carl D. Svard, P.E.
Dr. Robert V. Peltier, P.E., C.C.P.
James F. Ring
Michael P. Golden
Michael C. Loulakis, Esquire
Freeman Kirby
Gas Research Institute
Thomas E. Kalin
Donald O. Swenson
Rudy E. Theisen, P.E.
Winter Calvert, P.E.
Kellie Byerly
Anthony Pavone, P.E.
James R. Ross, P.E.
Robert J. Golden, Jr.
Luco R. DiNanno
David S. Milne, Jr.

Roger Yott, P.E.
Paul Gerst, P.E.
Sam A. Rushing
Dr. S. Somasundaram
Dr. M. Kevin Drost
Daryl R. Brown
Dr. Barney L. Capehart
Lynne C. Capehart, J.D.
William D. Orthwein
Milton Meckler, P.E., AIC
Terrence Kurtz
Ronald E. Russell, Esq.
Bernard F. Kolanowski
Martin Lensink, P.Eng.
Donald P. Galamaga, C.C.P.
Paul T. Bowen, P.E.
Rita Norton
Nayeem Sheikh
John Reader
Paul L. Multari
Jackie D. Moran
Terrence Kurtz

*Chapter 1*

# Cogeneration Basics—
# Theory and Practice

# Chapter 1

# Cogeneration Basics— Theory and Practice

*Richard D. Kennedy, P.E., C.E.M., C.C.P.*
*Citrus World*

I n the overall context of energy management theory, cogeneration is just another form of the conservation process. However, because of its potential for practical application to new or existing systems, it is carving a niche that is second to no other conservation technology.

This chapter explores the current theory and practice of cogeneration technology of single-stage steam back-pressure turbines, gas turbines and boiler applications.

---

Economic opportunities for cogeneration exist depending on the relative electrical and gas energy costs and the Public Service Commission regulations. In a growing number of cases, gas turbine cogeneration is a low risk, high return investment.

The first step in evaluating alternatives is to identify locations with a favorable cogeneration potential regardless of heat requirements. After selecting the most economically attractive location, the specific thermal requirements for the process or facility must be identified. Then, an analysis can be made to select the most attractive alternative for the specific application through computer-based economic models. Turbine field sales engineers can provide assistance in making an analysis.

The key evaluation factors can be classified by their relative impact as primary or secondary. The primary factors include:

- Value of electrical energy based on buy/sell rates.
- Fuel cost.
- Installed cost of the system.
- Utilization including an allowance for availability.

Secondary factors include:
- Net fuel rate (i.e., amount of fuel the cogeneration system uses, minus the fuel it saves).
- Maintenance costs (include spare parts consumption, overhaul costs and field service support).
- Operating personnel costs.
- Standby fees.

Accuracy of the primary factors is essential as a small error can significantly affect the results. An approximation of the secondary factors will not substantially affect the economic evaluation of the overall project.

Next to capital cost, the critical factors involved in determining the economics of a cogeneration unit are the fuel cost and the electrical rates faced by the prospective cogenerator. If one assumes that fuel and electricity vary at different rates, the evaluation may distort the economic analysis of cogeneration. If revenues (or savings) are projected to escalate at a higher rate than costs, even a poor investment decision may appear attractive.

The initial capital cost is critical in assessing the economic feasibility of a prospective cogeneration project since the length of the pay-back period is directly proportional to the initial capital cost.

It may be seen from Figure 1 that there are economies of scale that affect smaller combustion turbine cogeneration plants, particularly below 20 MW. In addition to higher unit capital costs ($/kW), smaller units also incur proportionately higher operation and maintenance costs. Each case must be analyzed on its own merits and parameters, however.

High capacity factors, although achievable, may be optimistic considering periodic outages that will occur for cogeneration plant maintenance, regular plant shutdowns that are scheduled in some industries, and hourly, daily, and weekly variations in steam demand that often occur. As the assumed capacity factor decreases, the economic payback period increases dramatically.

It may be concluded that ideal cogeneration applications should

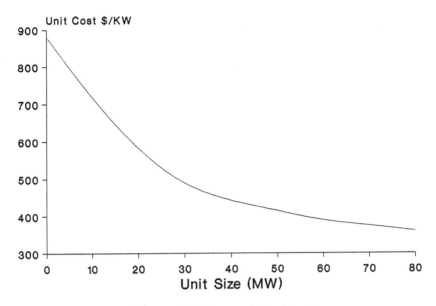

**Figure 1-1. Economies of Scale**

use the full output of the cogeneration unit on a 24-hour/day, 7-day/week basis. Lower usage factors will encounter more difficult economic hurdles. Therefore, cogeneration units serving seasonal thermal loads (such as for space heating) will, in many cases, be difficult to justify. A similar conclusion may be drawn for process applications that operate only one or two shifts per day.

The economics of cogeneration will be impacted by the reliability that is achieved for cogeneration units. First, if a cogeneration unit proves to be unreliable, its capacity factor will deteriorate, which, in turn, will adversely affect the economic payback for the investment. Also, poor unit reliability is associated with high maintenance costs and lengthy initial commissioning periods—either of which can eliminate the profitability originally perceived for a cogeneration project.

Perhaps even more critical is the potential disruption to a cogenerator's basic processes when an unplanned outage of the cogeneration unit occurs. To minimize such impact, cogeneration plants must be carefully engineered to provide appropriate backup steam supplies. Also, a combustion turbine plant is more complex than a conventional boiler, thus requiring a more sophisticated maintenance program.

Experience shows that achieving high reliability for a cogeneration plant requires not only a reliable combustion turbine, but also careful system engineering of balance-of-plant auxiliaries and support systems and an effective maintenance program.

There are two procedures for screening potential cogeneration applications discussed here. One is leasing and the other is purchase. The purchase approach is shown in Figure 1-2, which is based on these assumptions.

- Payback                    four years

- Capital cost               $700/kW-hour

- Net heat rate              6,400 Btu/kW-hour

- Maintenance cost           $0.0035/kW-hour

To use this figure, locate the point where the cogeneration fuel and utility electric costs intersect. If this point is above the anticipated annual utilization, the application will have a payback of less than 4 years showing a potential and should be studied further.

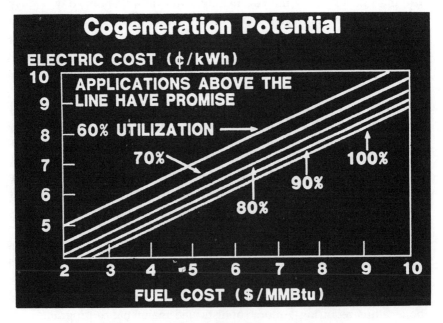

Figure 1-2. Purchase Approach to Screening

# COMMON ERRORS AND OMISSIONS

In the economic evaluation of a cogeneration system, some computer programs, worksheets and handbooks overlook the following factors that can affect the accuracy of the result:

1. **Heating Values**
   Utilities quote natural gas in "Higher Heating Values" (HHV) while manufacturers quote fuel consumed by gas turbines in "Lower Heating Values" (LHH). This amounts to about a 10 percent differential in cost per unit of natural gas.

2. **Services**
   Differences in quality, product development, service support, spare parts, and training can result in system availability ranging from a low of 60 percent to a high of 99 percent.

3. **Fuel Saved**
   If turbine exhaust is used to produce a given quantity of steam, the fuel savings equal to what would have been required to produce the same amount of steam in a fixed boiler should be credited to the system, rather than the amount of heat recovered from the turbine exhaust.

4. **Installation Credits**
   When a cogeneration system eliminates the need for boilers or emergency power generation equipment, the costs for these should be credited to the installation of the cogeneration system.

5. **Design Conditions**
   A cogeneration system is designed to ensure sufficient capacity at maximum ambient conditions. Since only a minimum amount of operating time is at these conditions, it would be inappropriate to use them in an economic evaluation.

6. **Part-Load Operation**
   To project realistic results, an adjustment must be made for a system not operating continuously at full load, such as the summer season of the citrus industry.

7. **Initial and Operating Costs**
   If costs for permits and standby fees are not included, the result will
   be an unrealistic evaluation.

8. **Risk Factors**
   An analysis of the impact made by the availability of fuel, the escalation of fuel and utility rates, and economic cycles should be included.

9. **Operating Modes**
   With varying requirements and controls, a different economic return
   will result from the alternative modes of operation which include
   heat-load following, electric-load following, and fixed electric load.
   The cost of power for each mode is required to determine the optimum mode for each operating condition and the appropriate control
   system.

   Two problems in the past have limited the economic feasibility of
cogeneration. First of all, previous installations were dependent on a
good energy balance. Suppose we design a cogeneration system which is
capable of simultaneously meeting the peak electrical and heat requirements.
   Then on reviewing the load cycle, we may discover that these peaks
do not occur simultaneously. In off-peak conditions, there may be substantial differences between the electrical and thermal requirements.
(Figure 1-3)
   If we have a high electrical demand and a low thermal demand, we
will be producing more heat than is required and will have to reject the
excess heat to the atmosphere.
   If we have a high thermal demand and low electrical demand, we
will have to add additional energy to the system. In either case, we will
be operating at off-optimum efficiency and the system economics will
suffer.
   Second, if we have a system demand as shown on the curve in
Figure 1-4, it would be advisable to select units equal in capacity to the
base load.
   To satisfy this load, we would require three units: one for the base
load capacity, a second for peaking capacity, and a third to serve as
standby for the other two.

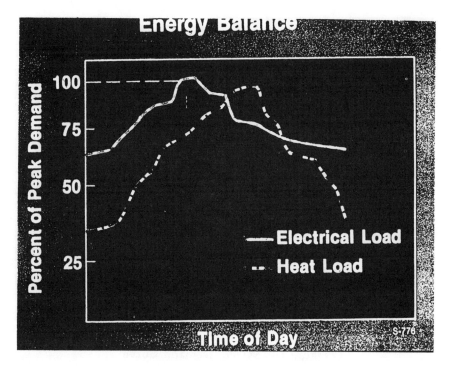

**Figure 1-3. Load Cycles**

A system like this would have an overall utilization of approximately 30 percent which would have to pay for the capital cost of a three-unit installation.

A solution to both of these problems would be to intertie the cogeneration system with the electric utility. We could then run the system to meet the heat load and use the electric utility grid to supply any storage or to absorb any excess power. Having the utility absorb excess power is not presently economical due to the present buy-back rates at this location (approximately $.02/kWh).

This system would be much less expensive and much simpler because we would require only one unit intertied with the utility. And the utilization would be nearly 100 percent for the example shown.

Most cogeneration systems require parallel operation with an electrical utility to effect a heat and electric load balance.

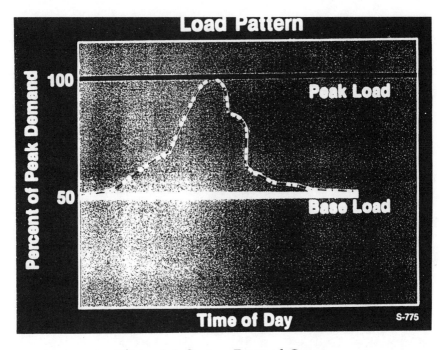

Figure 1-4. System Demand Curve

## ECONOMIC OPPORTUNITIES

Many factors are involved in evaluating a cogeneration system. However, many industrial and governmental facilities, domestic and foreign, are finding cogeneration to be a low risk, high return investment.

With spreadsheet programs such as Lotus 1-2-3, 20-20, Super Cal 4, etc., turbine or cogeneration representatives can give the prospective cogenerator a vast array of finished data to evaluate the economics of a site-specific turbine application.

Tables 1-1 and 1-2 are summary sheets and an economic analysis of our turbine project at Citrus World, Inc.

These are but two of hundreds of analyses that were run to evaluate the many variables that impact a cogeneration project. The economic analysis, Table 1-1, is a base case of 60,000 pounds of steam and 5.25¢/ kWh. It shows the effects of changing gas prices, a changing lease rate and a sensitivity analysis on additional gas burned in the duct burner over and above what the manufacturer specified concerning the operat-

## Table 1-1. Economic Analysis

Solar Centaur H Cogeneration System at Citrus World, Lake Wales Case 6C: 60,000 $/hr Average Steam Output & Elect. @ 5.25 cents/kWh

(BASE CASE)

Assumptions

| | | | | | | |
|---|---|---|---|---|---|---|
| A- | Average Steam Output (lb/hr): | | | 60,000 | | |
| B- | Electric Rate (cents/kWhr): | | | 5.25 | | |
| Gas Price (cents/therm): | | | 25 | 30 | 35 | 40 |
| C- | Generator Output (kW): | | | 3,500 | | |
| D- | Gas Turbine Fuel Flow (MMBtu/hr): | | | 45.2 | | |
| | Expected Duct Burner Fuel Usage (MMBtu/hr): | | | 39.1 | | |
| E- | Average Fuel Displaced (MMBtu/hr): | | | 75.0 | | |
| | Average Hours Operated/month: | | | 582 | | |
| | Year: | 1 | 2 | 3 | 4 | 5 |
| F- | Lease Payment ($/month): | $63,500 | $64,770 | $66,065 | $67,387 | $68,734 |

| Results ($1000/month) | Gas Price (c/therm, HHV) | | | |
|---|---|---|---|---|
| | 25 | 30 | 35 | 40 |
| Purchased Power Reduction | $106.9 | $106.9 | $106.9 | $106.9 |
| GT Fuel Cost | (72.3) | (86.8) | (101.3) | (115.7) |
| Nominal Duct Burner Fuel Cost | (62.6) | (75.1) | (87.6) | (100.1) |
| Steam Credit | 120.0 | 144.0 | 168.1 | 192.1 |
| | | | | |
| G - Gross Savings | $92.0 | $89.1 | $86.1 | $83.1 |
| Lease Payments (1st Year) | 63.5 | 63.5 | 63.5 | 63.5 |
| | | | | |
| Net Savings/month (1st year) | $28.5 | $25.6 | $22.6 | $19.6 |
| Net Savings/month (2nd year) | 27.3 | 24.3 | 21.3 | 18.3 |
| Net Savings/month (3rd year) | 26.0 | 23.0 | 20.0 | 17.0 |
| Net Savings/month (4th year) | 24.7 | 21.7 | 18.7 | 15.7 |
| Net Savings/month (5th year) | 23.3 | 20.3 | 17.4 | 14.4 |
| | | | | |
| Average Savings/Year | $311 | $276 | $240 | $204 |
| Total Savings over Lease Term | $1,557 | $1,378 | $1,200 | $1,021 |
| | | | | |
| Duct Burner Sensitivity Analysis: | | | | |
| 5% Additional DB Fuel Cost/Month | $3.1 | $3.8 | $4.4 | $5.0 |
| Net Savings/month (1st year) | $25.4 | $21.8 | $18.2 | $14.6 |
| Average Savings/year | $274 | $231 | $187 | $144 |
| Total Savings over Lease Term | $1,369 | $1,153 | $937 | $720 |
| | | | | |
| 10% Additional DB Fuel Cost/month | $6.3 | $7.5 | $8.8 | $10.0 |
| Net Savings/month (1st year) | $22.6 | $18.1 | $13.8 | $9.6 |
| Average Savings/year | $236 | $186 | $135 | $84 |
| Total Savings over Lease Term | $1,182 | $928 | $674 | $420 |

## Table 1-2. Economic Analysis - Summary Sheet

Average Savings/year ($1000)

| Case | Steam Output (#/hr) | Elect Rate (c/kWh) | Duct Burner Fuel Increase | Gas @ 25¢ | Gas @ 30¢ | Gas @ 35¢ | Gas @ 40¢ |
|------|------|------|------|------|------|------|------|
| 4A | 40,000 | 4.75 | 0% | $108 | $56 | 4 | ($48) |
|    |        |      | 5% | 90 | 35 | (21) | (76) |
|    |        |      | 10% | 73 | 14 | (45) | (104) |
| 4B | 40,000 | 5.00 | 0% | $169 | $117 | $65 | $13 |
|    |        |      | 5% | 152 | 96 | 40 | (15) |
|    |        |      | 10% | 134 | 75 | 16 | (43) |
| 4C | 40,000 | 5.25 | 0% | $230 | $156 | $126 | $74 |
|    |        |      | 5% | 213 | 157 | $126 | 46 |
|    |        |      | 10% | 195 | 136 | 77 | 18 |
| 4D | 40,000 | 5.50 | 0% | $298 | $187 | $187 | $135 |
|    |        |      | 5% | 274 | 218 | 163 | 107 |
|    |        |      | 10% | 256 | 197 | 138 | 79 |
| 5A | 50,000 | 4.75 | 0% | $149 | $105 | $61 | $17 |
|    |        |      | 5% | 121 | 72 | 22 | (27) |
|    |        |      | 10% | 93 | 39 | (16) | (71) |
| 5B | 50,000 | 5.00 | 0% | $210 | $166 | $222 | $78 |
|    |        |      | 5% | 182 | 133 | 83 | 34 |
|    |        |      | 10% | 155 | 100 | 45 | (10) |
| 5C | 50,000 | 5.25 | 0% | $271 | $227 | $183 | $139 |
|    |        |      | 5% | 243 | 194 | 144 | 95 |
|    |        |      | 10% | 216 | 161 | 106 | 51 |
| 5D | 50,000 | 5.50 | 0% | $332 | $288 | $244 | $200 |
|    |        |      | 5% | 304 | 255 | 206 | 156 |
|    |        |      | 10% | 277 | 222 | 167 | 112 |
| 6A | 60,000 | 4.75 | 0% | $189 | $153 | $118 | $82 |
|    |        |      | 5% | 152 | 108 | 65 | 22 |
|    |        |      | 10% | 114 | 63 | 13 | (38) |
| 6B | 60,000 | 5.00 | 0% | $250 | $215 | $179 | $143 |
|    |        |      | 5% | 213 | 170 | 126 | 83 |
|    |        |      | 10% | 175 | 124 | 74 | 23 |
| 6C* | 60,000 | 5.25 | 0% | $311 | $276 | $240 | $204 |
|    |        |      | 5% | 274 | 231 | 187 | 144 |
|    |        |      | 10% | 236 | 186 | 135 | 84 |
| 6D | 60,000 | 5.50 | 0% | $373 | $337 | $301 | $265 |
|    |        |      | 5% | 335 | 292 | 248 | 205 |
|    |        |      | 10% | 297 | 247 | 196 | 145 |
| 7A | 70,000 | 4.75 | 0% | $230 | $202 | $175 | $147 |
|    |        |      | 5% | 182 | 145 | 108 | 71 |
|    |        |      | 10% | 135 | 88 | 41 | (5) |
| 7B | 70,000 | 5.00 | 0% | $291 | $263 | $236 | $208 |
|    |        |      | 5% | 243 | 206 | 169 | 132 |
|    |        |      | 10% | 196 | 149 | 103 | 56 |
| 7C | 70,000 | 5.25 | 0% | $352 | $324 | $297 | $269 |
|    |        |      | 5 % | 305 | 267 | 230 | 193 |
|    |        |      | 10% | 257 | 210 | 164 | 117 |
| 7D | 70,000 | 5.50 | 0% | $413 | $386 | $358 | $330 |
|    |        |      | 5% | 366 | 328 | 291 | 254 |
|    |        |      | 10% | 318 | 271 | 225 | 178 |

*Base Case

ing parameters of the duct burner and boiler.

Table 1-2 is a summary of 192 different continuations of price and operating conditions, some of which are negative—basically costing more to run the project than income achieved.

Referring to Table 1-1, there are some items that require further explanation. Under assumptions, average steam output is in pounds/hour/averaged over a whole year.

Since most food processing plants normally have large changes in steam demand in season versus out of season, pounds/hour of steam calculated on a monthly basis in an off-season time frame will have substantially less steam produced than the in-season period. This is one of the items which will greatly affect the project savings. A month-by-month analysis is required to insure that the net savings does not become negative during these months. Even if a negative savings occurs, the project still could have merit if the total yearly savings can justify the system.

The electric rate (cents/kWh) needs to be carefully reviewed due to the effects of load factor. Mr. E. Coxe addresses this problem in a paper presented at an A.S.M.E. meeting.

*"Cogeneration will generally have an adverse effect on the average cost of any supplemental electricity purchased. This is a result of the load factor for purchased power decreasing as the amount of cogeneration is increased. The load factor is the ratio of the kilowatt-hours purchased during the month to the kilowatt-hours that would have been purchased had the requirements been at the billing demand for the entire month, or ratio of peak load to average load. This is illustrated in Table 1-3.*

*"The evaluation is made for generation levels of 0 to 75 percent of total kilowatt-hour requirements for a 1-month period. A standby charge has been added to cover the cost of the utility having to provide backup and other necessary services.*

*"Note that as the percent of cogeneration increases, the load factor for purchased electricity decreases from 67 percent to 33 percent. This has the effect of increasing the demand/standby component of the total bill from 20 percent to 40 percent. The overall effect is a 34 percent increase in the per-unit cost of purchased electricity compared to that with no cogeneration. Although the savings can still be substantial, this represents a reduction in the expected revenues.*

*"There are several options for the power generated in a cogeneration facility. The first is to generate at a level that never exceeds what can be utilized on-site and purchase any additional requirements from the utility. In addition to supplemental power requirements, the utility would provide all standby and backup power at a rate approved by the Florida Public Service Commission (PSC).*

*"The second is to generate at a level exceeding your requirements and sell any excess power to the utility. The revenue from the sale of power would have to exceed the fuel and maintenance costs associated with the additional generation for it to be attractive.*

*"The last is to sell all of the cogenerated power to the utility and purchase all of the facility power requirements from the utility at the current general service rate schedule. There would be no backup or standby charges other than those inherent in the general service schedule. This approach is, in general, not as attractive as that of offsetting purchased power."*

Generator output in kilowatts is variable on a single-shaft machine. The Centaur "H" machine, while only rated at 3.5 MW (72°F), will produce 4.0 MW on a normal cold winter day in Florida during January and February. The opposite is also true in the summer months and the Centaur "H" will only produce 3.2 to 3.3 MW on a hot summer day, July to September. Again, a detailed month-by-month analysis is the key to a successful project.

The economics of cogeneration units are sensitive to fuel prices and electric rates, which obviously vary from region to region. For typical fuel and electric costs prevalent in many areas of the United States, the larger combustion turbine cogeneration units have attractive after-tax payback periods.

Cogeneration economics deteriorate rapidly as the unit capacity factor decreases. This will discourage cogeneration units for applications that do not require process heat on an essentially continuous around-the-clock basis.

If electric rates change faster than fuel prices (particularly natural gas), then the economic feasibility of cogeneration will be enhanced. However, differential rates may distort the projected payback period and, therefore, any such assumption should be carefully justified.

There are continuing uncertainties involved in establishing capacity credits for electric power sold by cogenerators to utilities. In some areas,

## Table 1-3. Effect of Demand and Standby Charges on Supplemental Power Cost

| Demand | = | 12000 kW | Cogenerations | | |
|---|---|---|---|---|---|
| Consumption | = | 5670000 kWh | Energy Charge | = | $0.05/kW |
| Average | = | 8000 kW | Demand Charge | = | $4.50/kW |
| Energy Charge | = | $0.05/kWh | Standby Charge | = | $2.00/kW |
| Demand Charge | = | $6.50/kW | Cogen Fuel/O&M | = | $0.02/kWh |

| Percent Generation | Generation kW | Purchased kWh | Demand kW | Load Factor | Energy Cost | Demand Charge | Standby Charge (1) | Total Cost | Cents per kWh | Percent Increase |
|---|---|---|---|---|---|---|---|---|---|---|
| 0% | 0 | 5670000 | 12000 | 66.7% | 288000 | 78000 | 0 | 366000 | 0./0635 | 0 |
| 25% | 2000 | 4320000 | 10000 | 60.0% | 216000 | 45000 | 24000 | 285000 | 0.0660 | 3.81 |
| 50% | 4000 | 2880000 | 8000 | 50.0% | 144000 | 36000 | 24000 | 204000 | 0.0708 | 1 1.5% |
| 75% | 6000 | 1440000 | 6000 | 33.3% | 72000 | 27000 | 24000 | 123000 | 0.0854 | 34.4% |

Notes: (1) Calculated based on the Contract Demand, i.e., the maximum level of standby required.

| Percent Generation | Generation kWh | Cost @ $02/kWh (2) | Purchased kWh | Cost @ $.0635/kWh | Actual Cost | Expected Savings | Actual Savings | Monthly Difference | Annual Difference |
|---|---|---|---|---|---|---|---|---|---|
| 0% | 0 | 0 | 5670000 | 366000 | 366000 | 0 | 0 | 0 | 0 |
| 25% | 1440000 | 28800 | 4320000 | 274500 | 285000 | 62700 | 52200 | 10500 | 126000 |
| 50% | 2880000 | 57600 | 2880000 | 183000 | 204000 | 125400 | 104400 | 21000 | 252000 |
| 75% | 4320000 | 86400 | 1440000 | 91500 | 123000 | 188100 | 156600 | 31500 | 378000 |

Notes: (2) Cost based on displacing fuel at $2.80 per million Btu burned in an existing boiler at 73% efficiency.

Gas Turbine O & M at 0.35 cents per kWh.

this will tend to discourage prospective cogenerators that cannot use the total electrical output of a cogeneration unit.

Innovative financing and ownership concepts will continue to play a prominent role but can add complications and potential project delays compared with conventional arrangements.

Cogeneration projects that "seem good on paper" can encounter practical difficulties that eliminate the profitability of the project. To minimize the prospect for such problems, thorough engineering of each cogeneration project will be required.

There are several small to large turbine manufacturers in the market place at the present time. Many offer some form of waste heat packaging and, with the electronic spreadsheet as described, the economics should be investigated as to the economic fit to the processing plant.

Some of these combinations offer very good heat rates and efficiencies. In Table 1-4, the calculations are shown for a medium processor operating 7 months in-season and 5 months off-season. Remembering that this example is a weighted average, it becomes apparent that a processing plant operating at this level reaches an operating level where it is fuel/price insensitive. Basically, the use of fuel offsets the fuel displaced. Operating above this level displaces more fuel than consumed, the result is increased profit or revenue at any price of gas; the higher the price of fuel, the greater the profit margin.

This unique combination again points out the need for a very detailed accounting of how the cogeneration system is to be operated over the entire year.

## INSTALLATION

When the decision is made to proceed with a cogeneration project, the installation and start-up phases are just as important as the economic analysis. One of the first decisions is whether to make it a turnkey installation where the turbine company or some outside corporation handles the construction and start-up, or be handled internally by the company or even some combination of the two.

The turnkey decision places the bulk of the technical, delivery and construction details with an outside firm for a negotiated cost. The quality of the completed job is only as good as the contractor doing the installation. The contract should have as much detail as possible to in-

### Table 1-4. Calculation of Weighted Average Values
### For 110,000 lb/hr Waste Heat Boiler Operating @100,000 lb
### (Breakdown Fuel Consumption Scenario)

|  | On-Season | Off-Season | Total | Weighted Average |
|---|---|---|---|---|
| Duration (months) | 7 |  | 5 |  |
| Total hours in month | 730 | 730 |  |  |
| Utilization (days/week) | 6 | 5 |  |  |
| Net Operating Hours/month | 626 | 521 |  | 582 |
| Total Operating Hours/year | 4,380 | 2,607 | 6,987 | 79.8% |
|  |  |  |  |  |
| Avg Steam Output (#/hr) | 100,000 | 30,000 |  | 73,881 |
| Total Steam Output (1,000#/Yr) | 438,000 | 78,214 | 516,214 |  |
|  |  |  |  |  |
| Supply Fuel (MMBtu/hr) | 80.83 | 7.95 |  | 53.6 |
| Total Supply Fuel (MMBtu/yr) | 354,035 | 20,727 | 374,762 |  |
|  |  |  |  |  |
| Displaced Boiler Efficiency | 75% | 75% |  |  |
| Feedwater Temperature (deg F) | 226 | 226 |  |  |
| Heat Required (Btu/lb) | 1003 | 1003 |  |  |
| Fuel Displaced (MMBtu/hr) | 133.7 | 40.1 |  | 98.8 |
| Total Fuel Displaced (MMBtu/yr) | 585,752 | 104,599 | 690,351 |  |
|  |  |  |  |  |
| Gas Turbine Fuel Flow (MMBtu/hr) | 45.2 | 45.2 |  | 45.2 |
| Supply Fuel (MMBtu/hr) | 80.8 | 8 0 | _____ | 53.6 |
| Total Cogeneration Fuel (MMBtu/hr) | 126.0 | 53.2 |  | 98.8 |

sure that the installation fits the needs of the company using the cogeneration facility.

The opposite is the situation where the owner handles the entire project. If the owner has a good technical staff, this will usually give a better installation, and one that better fits the actual operating needs of the plant requiring steam and electrical energy.

An example is steam supply to a food processing facility. The cogenerator will pay a standby charge as part of the rate schedule for generating electrical energy. If the turbine tripped, load would be assumed by the utility without an interruption to service.

This is not the case when it comes to steam supply for the processing plant. Most processors require uninterrupted steam supply in season. This can be accomplished by adding a fresh-air fan to the boiler system which will permit independent boiler operation if the turbine is off-line for any reason.

Backup is required if there is only one turbine/boiler installation, and boiler downtime will be needed for maintenance, inspection and minor or major repairs. A multi-turbine/boiler installation probably would not need additional standby boiler capacity if one unit could be off-line and plant needs supported from the remaining on-line boiler.

In summary, there are substantial opportunities for cogeneration. The economics of both electric and steam need to be identified and evaluated. With the importance of changing fuel costs, the contribution of the thermal savings from a cogeneration project can greatly outweigh the savings from electricity alone and the impact of both needs to be considered in any cogeneration project. To realize the projected savings, all the elements that can be identified through input from the owner and the turbine and boiler manufacturers need to be assessed and evaluated against how the system is to be actually run month by month. Electronic spread sheets now make this a realistic approach to assessing a cogeneration facility.

*Chapter 2*

# Cogeneration Systems Analysis

*Chapter 2*

# Cogeneration Systems Analysis

*Cecelio E. Gracias and Lloyd S. Hoffstatter*
*New York State Energy Office*

ogeneration is the process whereby the Energy User generates its own electricity to satisfy in whole or in part the facility's electrical power needs and simultaneously recovers the otherwise wasted heat of combustion in an attempt to meet at least part of the facility's thermal load.

Cogeneration is cost effective only if there is a thermal load which fairly matches or exceeds the heat output of the cogenerator. Ideally, the cogeneration thermal output should be able to satisfy all of the existing thermal load, but the condition is not necessary as long as there is enough usable recovery to yield an acceptable payback.

The prime mover in a cogenerating unit is either a fossil fuel-fired engine, or a gas turbine or a steam turbine powered with steam from a primary or a waste heat boiler (viz. Combined Cycle Systems). For most facilities, an engine powered by direct combustion of gas or oil is the most common solution.

The general requisites for **Cogeneration** are: (1) a <u>high electric demand</u> coupled with heavy electric energy consumption; (2) a suitably <u>large thermal load</u> (for heating or process or both); (3) <u>high demand and/or energy unit costs</u>. **Cost effectiveness** results from a suitable combination of these three factors. Yet, a very high demand lessens the effect of a high unit cost of demand when cogeneration is implemented. The following points should be considered when evaluating cogeneration:

1.    Cogeneration to meet demand primarily;

2.    Cogeneration to meet electrical consumption;
3.    Cogeneration to meet thermal load primarily.

Cogenerating systems designed to meet peak demand alone are not cost effective. Viable alternatives would be: a) to consider meeting average demand, with occasional excess capacity; or b) to accept a lower cogenerating capacity to just satisfy energy requirements, paying the possible penalties of frequent utility backup and dependency on existing boilers to satisfy residual thermal load.

There is a further consideration which may be the overriding factor in deciding the final cogenerating configuration—that of availability of the appropriate cogeneration engine fuel. Once the main objective for installing cogeneration is established, the selection of the best operating strategy should be based on site load duration curves.

Following are potentially effective operating strategies:

a.    "Base Load" thermally and electrically (continuing to buy power and make steam);
b.    "Follow" thermal load, displace part of purchased power (idle boilers, purchase supplemental power as required);
c.    "Follow" electrical load, displace a portion of the thermal load (minimum purchase: standby only);
d.    "Follow" electrical load, idle all boilers and supplement steam demand with extraction condensing combined cycle;
e.    "PURPA ENGINE"—generate power for sale and buy steam.

Once the cogeneration system is installed, use the load duration curves to match generation and heat recovery for optimum utilization of cogeneration capacity.

# N.Y. STATE PROPOSED COGEN SYSTEMS

Through the DOE ICP program, potential cogeneration situations are first analyzed in the form of Technical Assistance Studies (TAS) and may be submitted concurrently for funding in a concurrent or subsequent Energy Conservation Measures (ECM) Application. These Cogeneration Proposals have been filed with the N.Y. State Energy Office under the ICP program under Cycles XIII, XIV, and XV:

1. S.U.N.Y.[1] Stony Brook, NY (Mar/81) - 2×16.3MW[2]   PB = 6.3 yrs
2. S.U.N.Y. Oneonta, NY (Apr/81) - 3400 kW   PB = 7.1 yrs
3. S.U.N.Y., Oswego, NY (Apr/81) - 4300 kW   PB = 7.2 yrs
4. S.U.N.Y., Brockport, NY (Apr/81) - 6000 kW   PB = 7.2 yrs
5. Creedmore Psychiatric Center, Queens Village, NY (Oct/81) - 1410 kW   PB = 3.0 yrs
6. Yeshiva Flatbush High (Mar/89) - 60 kW   PB = 10.4 yrs
7. Long Island College Hospital (Feb/89) - 1200 kW   PB = 2.7 yrs
8. Lutheran Medical Center, Brooklyn, NY- 2×860 kW   PB = 2.5 yrs
9. Albany Memorial Hospital, Albany, NY - 525 kW   PB = 3.8 yrs
10. Montefiore Hospital (Feb/89) - 3200 kW   PB = 2.8 yrs
11. St. Johns' Riverside Hospital, Yonkers, NY (Apr/89) -1260 kW   PB = 5.4 yrs
12. St. Joseph's Med. Ctr. (Oct/88) - 350K   PB = 11.4 yrs
13. Waverly CSD (Nov/89) - 300 kW   PB = N/A
14. St. Barnabas Hospital, Bronx, NY (Dec/89) - 1000 kW   PB = 2.4 yrs
15. Bayley Seton Hospital (Apr/90) - 548 kW   PB = 3.9 yrs
16. Phelps Memorial Hospital (Apr/90) - 2×350 kW   PB = 3.7 yrs
17. Calvary Hospital (1993) - 450 kW   PB = 4.3 yrs
18. Catholic Med. Ctr. St. Johns Hosp. (1993) - 600 kW   PB = 3.9 yrs
19. Saratoga Hospital (1993) - 515 kW   PB = 4.3 yrs
20. Carthage High School (1993) - 72 kW   PB = 5.5 yrs

# ENERGY BALANCE IN COGENERATION

Consider a cogeneration system consisting of the prime mover connected to an electrical generator. The heat engine may be a gas turbine, a steam-operated turbine or a reciprocating engine. Thermodynamically, the energy balance in the cogeneration system can be defined in terms of the following relationships:

**Available Heat** = Heat of Combustion - Exhaust Losses
**Available Power** = Available Heat - Engine Losses
**Shaft Power** = Available Power - Derating Losses
**Cogenerated Electric Power** = Shaft Power - Generator Losses

All of the above quantities can be calculated, but unfortunately input data are not readily available. Instead, the designer uses the performance curves provided by the manufacturer.


# METHODOLOGY

### The Cogeneration Process

Before going into the details of the spreadsheet construction, a review of the cogeneration process is in order. By definition, *cogeneration* is the process by which electricity is generated on site by the user to displace electrical energy purchased from the utility. Cogeneration is economic only when part of the existing thermal load can be replaced by the recovery of the heat contained in the effluent gas. The following quantities are then essential in characterizing the performance of the proposed cogeneration system:

$P_c$ = cogenerating capacity (kW)
$Q_c$ = rate of fuel combustion = $Q_f {}^* R_f$ (Btuh)
$Q_a$ = heat rate (Btuh/kW) or available fuel heat/kW
$Q_r$ = exhaust heat recovery rate (Btu/kWh)
$H_c$ = number of cogeneration hours (H/Y)
$K_e$ = 3142 Btuh/kW
$Q_w$ = unrecovered exhaust heat (Btuh/kW)
$Q_f$ = Btu per unit fuel
$R_f$ = units of fuel per hour (Btu/Gal or Therm)

In a practical application, $Q_a$ for a given cogeneration unit is taken from the manufacturer's data sheet for a desired value of $P_c$ the cogeneration capacity of the unit. Otherwise, $Q_a$ would have to be calculated from $Q_c$ using the fuel heating value $Q_f$ and its consumption rate $R_f$. Considerations of energy balance would then lead to the following relationships:

For the thermal process: $Q_a = Q_r + Q_w = Q_f/\varepsilon_h$

For the generation process: $Q_r = P_c {}^* H_c {}^* K_e/\varepsilon_e$

For the overall process:

$$Q_r = P_c{}^*H_c{}^*K_e/(\varepsilon_e{}^*\varepsilon_h) = P_c{}^*H_c{}^*K_e/\varepsilon_c$$

$\varepsilon_c$= cogeneration efficiency = $\varepsilon_e{}^*\varepsilon_h$

where $\varepsilon_e$ and $\varepsilon_h$ are the efficiencies for the electric and thermal conversion processes, respectively.

### The Cogeneration Template

In its present state, the N.Y. State Energy Office Template allows a quick evaluation of cogeneration potential in a wide variety of situations or any given capacity of Cogenerator. It will analyze cost effectiveness of full or part time cogeneration with mixed fuels for both Boiler and Cogenerator based on an average or period fuel costs.[3] The attached Sample (Table 2-1) shows a blank Cogeneration Spreadsheet.

The Template consists of three main Sections:

1. The <u>Data Display Section</u>, situated at the left-most area of the spreadsheet; this section comprises three tables which respectively contain: the Fuel Data; the Cogeneration Data; and the Calculated Parameters.

2. The <u>Fuel Consumption Section</u>, which occupies the top of the spreadsheet. This section contains all the Utilities data, namely quantity of fuel used and respective costs for the indicated periods (usually monthly). Because this section represents a reading table for the computational section, care must be exercised to input the data in the allocated slots.

3. The <u>Computational Section</u>, which contains all the formulae needed to calculate the various intermediates quantities required to estimate net savings from cogeneration and the simple payback over the period selected for analysis. The computational section contains the following: the calculated energy data; the estimated costs of fuel; the projected energy and energy cost savings; and a summary table, which among other quantities, shows the simple payback for the period.

# Table 2-1. Cogeneration Template (Worktable)

COGENERATION ANALYSIS

**FINAL EDITION**

**FUEL DATA**

| BOILER FUEL: | Fraction |
|---|---|
| Natural Gas | 0.00 |
| Other Fuel | 1.00 |
| COGEN FUEL: | |
| Natural Gas | 1.00 |
| Other Fuel | 0.00 |

**INPUT DATA**

| | |
|---|---|
| KW rating | 0 |
| Heat rate, BTU/KWH | 0 |
| #HRS/Day cogen. | 0 |
| Exh.recov.,BTU/KWH | 0 |
| BTU/KWH, Source | 11600 |
| Boiler, Efficiency | 0 |
| BTU/UnitBoilNGas | 0 |
| BTU/UnitBoilFOil | 0 |
| BTU/UnitCogNGas | 0 |
| $Cost/UnitCogNGas | 0 |
| BTU/UnitCogFOil | 0 |
| $Cost/UnitCogFOil | 0 |
| CogenInst. Cost | 0 |

**CALCULATED PARAMETERS**

| | |
|---|---|
| El. Cost, $/KWH | $0.0000 |
| $Cost/MBTUBoilNGas | $0.000 |
| $Cost/MBtuBoilFOil | $0.000 |
| $Cost/MBTUCogNGas | $0.000 |
| $Cost/MBTUCogFOil | $0.000 |
| CogThermEff. | 0.000 |
| CogElecEff. | 0.000 |
| TotCogEff. | 0.000 |
| $Cost/KWCog | $0 |

| # | | Jan | Feb | Mar | Apr |
|---|---|---|---|---|---|
| 1 | Month ---> | Jan | Feb | Mar | Apr |
| 2 | *Utility Billings* | | | | |
| 3 | No. days    0 | 0 | 0 | 0 | 0 |
| 4 | KW demand | 0 | 0 | 0 | 0 |
| 5 | KWH billed | 0 | 0 | 0 | 0 |
| 6 | $Cost billed | 0 | 0 | 0 | 0 |
| 7 | No. days | 0 | 0 | 0 | 0 |
| 8 | #therms billed | 0 | 0 | 0 | 0 |
| 9 | $Cost billed | 0 | 0 | 0 | 0 |
| 10 | #Gal billed | 0 | 0 | 0 | 0 |
| 11 | $Cost billed | 0 | 0 | 0 | 0 |
| 12 | *Energy Data* | | | | |
| 13 | BoilerFuel,MBTU | 0 | 0 | 0 | 0 |
| 14 | EngFuel, MBTU | 0 | 0 | 0 | 0 |
| 15 | Tot.Ht.avail, MBTU | 0 | 0 | 0 | 0 |
| 16 | Avoided fuel, MBTU | 0 | 0 | 0 | 0 |
| 17 | NetBoil.fuel, MBTU | 0 | 0 | 0 | 0 |
| 18 | Cog+BoilFuel, MBTU | 0 | 0 | 0 | 0 |
| 19 | Max Cogen KWH | 0 | 0 | 0 | 0 |
| 20 | KWH displ.Cogen. | 0 | 0 | 0 | 0 |
| 21 | *Costs* | | | | |
| 22 | $Cost Eng. Fuel | $0 | $0 | $0 | $0 |
| 23 | $Cost NetBoil.Fuel | $0 | $0 | $0 | $0 |
| 24 | $TotCost w/Cogen. | $0 | $0 | $0 | $0 |
| 25 | *Savings* | | | | |
| 26 | TotElec. Savings | $0 | $0 | $0 | $0 |
| 27 | Thermal Savings | $0 | $0 | $0 | $0 |
| 28 | NetSav from Cogen | $0 | $0 | $0 | $0 |
| 29 | NetMBTUSav,Source | 0 | 0 | 0 | 0 |
| 30 | | | | | |
| 31 | *Summary:* | SourceRelSavings,MBTU = | | | 0 |
| 32 | | | | | |
| 33 | PERIOD-BASED COSTS: | | | | |
| 34 | $Cost/KWH | $0 | $0 | $0 | $0 |
| 35 | $/UnitBoilGas | $0 | $0 | $0 | $0 |
| 36 | $/UnitBoilFOil | $0 | $0 | $0 | $0 |
| 37 | $CostBoilFuel | $0 | $0 | $0 | $0 |
| 38 | $CostEng.Fuel | $0 | $0 | $0 | $0 |
| 39 | $CostAvoidedFuel | $0 | $0 | $0 | $0 |
| 40 | $CostNetBoilFuel | $0 | $0 | $0 | $0 |
| 41 | $TotCost w/Cogen | $0 | $0 | $0 | $0 |
| 42 | Boil.FuelSavings | $0 | $0 | $0 | $0 |
| 43 | TotElecSavings | $0 | $0 | $0 | $0 |
| 44 | NetSavfrom Cogen | $0 | $0 | $0 | $0 |
| 45 | | | | | |
| 46 | | | | | |
| 47 | | | | | |
| 48 | | | | | |
| 49 | | | | | |
| 50 | | | | | |
| 51 | | | | | |
| 52 | | | | | |

| May | Jun | Jul | Aug | Sep | Oct | Nov | Dec | Total |
|---|---|---|---|---|---|---|---|---|
| 0 | 0 | 0 | 0 | 0 | 0 | 0 | 0 | 0 |
| 0 | 0 | 0 | 0 | 0 | 0 | 0 | 0 | 0 |
| 0 | 0 | 0 | 0 | 0 | 0 | 0 | 0 | 0 |
| 0 | 0 | 0 | 0 | 0 | 0 | 0 | 0 | $0 |
| 0 | 0 | 0 | 0 | 0 | 0 | 0 | 0 | 0 |
| 0 | 0 | 0 | 0 | 0 | 0 | 0 | 0 | 0 |
| 0 | 0 | 0 | 0 | 0 | 0 | 0 | 0 | $0 |
| 0 | 0 | 0 | 0 | 0 | 0 | 0 | 0 | 0 |
| 0 | 0 | 0 | 0 | 0 | 0 | 0 | 0 | $0 |

Total

| | | | | | | | | |
|---|---|---|---|---|---|---|---|---|
| 0 | 0 | 0 | 0 | 0 | 0 | 0 | 0 | 0 |
| 0 | 0 | 0 | 0 | 0 | 0 | 0 | 0 | 0 |
| 0 | 0 | 0 | 0 | 0 | 0 | 0 | 0 | 0 |
| 0 | 0 | 0 | 0 | 0 | 0 | 0 | 0 | 0 |
| 0 | 0 | 0 | 0 | 0 | 0 | 0 | 0 | 0 |
| 0 | 0 | 0 | 0 | 0 | 0 | 0 | 0 | 0 |
| 0 | 0 | 0 | 0 | 0 | 0 | 0 | 0 | 0 |
| 0 | 0 | 0 | 0 | 0 | 0 | 0 | 0 | 0 |

| $0 | $0 | $0 | $0 | $0 | $0 | $0 | $0 | $0 |
|---|---|---|---|---|---|---|---|---|
| $0 | $0 | $0 | $0 | $0 | $0 | $0 | $0 | $0 |
| $0 | $0 | $0 | $0 | $0 | $0 | $0 | $0 | $0 |

| $0 | $0 | $0 | $0 | $0 | $0 | $0 | $0 | $0 |
|---|---|---|---|---|---|---|---|---|
| $0 | $0 | $0 | $0 | $0 | $0 | $0 | $0 | $0 |
| $0 | $0 | $0 | $0 | $0 | $0 | $0 | $0 | $0 |
| 0 | 0 | 0 | 0 | 0 | 0 | 0 | 0 | 0 |

$InstCost/KWCog = $0        Payback = 0.00 Yrs

| $0 | $0 | $0 | $0 | $0 | $0 | $0 | $0 | $0 |
|---|---|---|---|---|---|---|---|---|
| $0 | $0 | $0 | $0 | $0 | $0 | $0 | $0 | $0 |
| $0 | $0 | $0 | $0 | $0 | $0 | $0 | $0 | $0 |
| $0 | $0 | $0 | $0 | $0 | $0 | $0 | $0 | $0 |
| $0 | $0 | $0 | $0 | $0 | $0 | $0 | $0 | $0 |
| $0 | $0 | $0 | $0 | $0 | $0 | $0 | $0 | $0 |
| $0 | $0 | $0 | $0 | $0 | $0 | $0 | $0 | $0 |
| $0 | $0 | $0 | $0 | $0 | $0 | $0 | $0 | $0 |
| $0 | $0 | $0 | $0 | $0 | $0 | $0 | $0 | $0 |
| $0 | $0 | $0 | $0 | $0 | $0 | $0 | $0 | $0 |
| $0 | $0 | $0 | $0 | $0 | $0 | $0 | $0 | $0 |

Payback = 0.00 Yrs

The usual period for analysis is that of twelve consecutive months of operation. The last column on the Fuel Consumption Section shows the total over this period. From this column the annualized consumption data and fuel unit costs may be calculated, as shown in the Calculated Parameters section. The latter also includes the calculated values for the various efficiencies which may be defined for the cogeneration system (i.e., thermal, electrical and overall). These calculated efficiencies usually show a close agreement with values reported elsewhere (e.g., TA Studies).

Although for NYSEO purposes the use of average cost is sufficient for the purposes of the ICP program, period-based costs are also included in the section below the Summary of Results line. The latter are more useful for assessing cogeneration needs for specific periods (e.g., for seasonal variations) and later on to control operation of the installed cogeneration system.

## Input Data

These refer to the information contained in the Input Table and consists of design data needed to define the proposed cogeneration system(s). They are: the contemplated *capacity* (kW), the cogeneration fuel, and the *operating schedule* (# hours/year), with indication of fuel types used. Also included is the information on the heating system whose output will be, at least in part, replaced by the cogenerating system thermal output. In the case of dual fuel systems, an estimate or assumption of the relative consumption is made by assigning consumption fractions (listed in the Fuel Data Table).

Besides cogenerating capacity, the following additional information is needed: the *heat rate* (Btuh/kW or Btu/kWh), which describes the thermal energy released by combustion of the cogeneration fuel; and the *exhaust heat recovery rate* (Btu/kWh), which expresses the amount of heat available for displacement of current heating load.

## Consumption Data

These refer to the information derived from the Utilities bills and constitute the Fuel Consumption Table. As mentioned, the Fuel Data Table allows for the often encountered situation in which two fuels are alternately used to accommodate temporary scarcity or to compensate for fluctuations in fuel allocation (as is often found in natural gas supply). Once the two Input Tables are filled with the appropriate informa-

tion the Template quickly calculates the following quantities
for a variety of fuel utilization ratios:

1. The source-related savings (MBtu/yr);
2. The cost per kW cogenerated; and
3. The project simple payback period.
4. The cogeneration thermal output efficiency,
5. Cogeneration electric output efficiency,
6. Cogeneration overall efficiency.

Also calculated are MBtu costs of boiler fuels, cogenerator fuels and
billed electricity, as well as engine fuel costs and thermal and electric cost
savings.

**Analysis of Cogeneration Systems**
An application, using a single fuel scenario, is illustrated by Table
2-2. This example corresponds to the following hypothetical situation: A
large building complex consumes electricity for lighting, refrigeration
and air conditioning. Peak demand is 1944 kW (January) and the low
occurs at 1224 kW (April), encompassing a span of 720 kW. Space heating requirements are satisfied by several boilers burning #4 oil, occasionally supplemented by LPG (propane).

The utility bills are:

|  | ELECTRICITY | #4 F.O. |
|---|---|---|
| QUANTITY | 7,369,200 kWh | 288,253 Gal. |
| COST | $523,772 | $194, 128 |
| UNIT COST | $0.0711/kWh | $0.673/Gal. |

**"What If" Analysis**
Once the Consumption and Input Data are typed into the Template, the values of the remaining cells will be quickly and automatically
calculated. Because of this particular feature of the Template, **what if
analyses** may be carried out by varying the Cogeneration parameters,
the Utilization factors[4] and any other input parameters such as daily
operating hours, fuel unit costs, etc. This type of analysis is the most
valuable feature of the template application to cogeneration analysis as
discussed in the next section.

## Table 2-2. Cogeneration Analysis: Summary
## (Fuel = Fuel Oil)

| EngineSize: kW | 100 | 200 | 300 | 400 | 500 | 600 |
|---|---|---|---|---|---|---|
| Eng Fuel, MBtu | 12080 | 24160 | 36240 | 48320 | 60400 | 72480 |
| Ht. avail, MBtu | 3066 | 6132 | 9198 | 12264 | 15330 | 18396 |
| Max Cogen kWh | 876000 | 1752000 | 2628000 | 3504000 | 4380000 | 5256000 |
| kWh displ Cog | 876000 | 1752000 | 2628000 | 3504000 | 4380000 | 5256000 |
| % kWh displ | 12.13 | 24.26 | 36.39 | 48.53 | 60.66 | 72.79 |
| $Eng Fuel (G) | 62114 | 124227 | 186341 | 248455 | 310568 | 372682 |
| $Net Boil (FO/LPG) | 871844 | 815290 | 758735 | 702181 | 645627 | 589072 |
| $Tot Cost Fuel | 933957.6 | 939517 | 945076.3 | 950635.7 | 956195 | 961754.4 |
| $Net Sav Coge | 56703 | 113406 | 170109 | 226812 | 283515 | 340218 |
| 10* CCI | 0.607127 | 1.207069 | 1.799952 | 2.385901 | 2.965037 | 3.537477 |
| CCI-Avg | 0.060713 | 0.120707 | 0.179995 | 0.23859 | 0.296504 | 0.353748 |
| FUR—Avg | 0.5648 | 0.5648 | 0 5648 | 0.5648 | 0.5648 | 0.5648 |
| TLM - Avg | 0.2492 | 0.4983 | 0.7475 | 0.9966 | 1.2458 | 1.4949 |
| Payback | 3.53 | 3.53 | 3.53 | 3.53 | 3.53 | 3.53 |
| 10* FUR | 5.648 | 5.648 | 5.648 | 5.648 | 5.648 | 5.648 |
| $Eng Fu/kWh | 0.0709 | 0.0709 | 0.0709 | 0.0709 | 0.0709 | 0.0709 |
| $Tot Fuel/kWh | 13.1244 | 11.8313 | 10.7815 | 9.9123 | 9.1807 | 8.5566 |
| Engine Size: kW | 100 | 200 | 300 - | 400 | 500 | 600 |

## Cogeneration Cost Analysis

To determine the most effective cogeneration engine size, the Template was used to analyze the costs of cogeneration for two different fuels—oil and natural gas over a capacity ranging from 100 kW to 1500 kW. Lack of space does not permit the displaying of the cost parameters versus size. However, analysis of this type of plot reveals that the optimum size should occur, when the following condition is verified:

$$\$ \text{ Eng Fuel} = \$ \text{ Net Boiler Fuel}$$

However, if one were to draw a plot of savings vs. size, the curve will show that maximum savings occurs right near the point where above condition is satisfied. Tables 2-2 and 2-3 contain the data necessary

| 700 | 800 | 900 | 1000 | 1100 | 1200 | 1300 | 1400 | 1500 |
|---|---|---|---|---|---|---|---|---|
| 84560 | 96640 | 108720 | 120800 | 132880 | 144960 | 157041 | 169121 | 181201 |
| 21462 | 24528 | 27594 | 30660 | 33726 | 36792 | 39858 | 42924 | 45990 |
| 6132000 | 7008000 | 7884000 | 8760000 | 9836000 | 10512000 | 11388000 | 12264000 | 13140000 |
| 6062400 | 6762000 | 7196400 | 7323600 | 7369200 | 7369200 | 7369200 | 7369200 | 7369200 |
| 83.62 | 92.69 | 98.08 | 99.49 | 100.00 | 100.00 | 100.00 | 100.00 | 100.00 |
| 434796 | 496909 | 559023 | 621137 | 683250 | 745364 | 807478 | 869592 | 931705 |
| 534148 | 487200 | 443092 | 401102 | 362775 | 325279 | 293022 | 265687 | 242446 |
| 968943.6 | 984109.4 | 1002115 | 1022239 | 104606 | 1070643 | 1100499 | 1135279 | 1174151 |
| 390345 | 424904 | 437773 | 426690 | 406145 | 381527 | 351671 | 316891 | 278019 |
| 4.028561 | 4.317646 | 4.368492 | 4.174072 | 3.882741 | 3.563534 | 3.195557 | 2.791309 | 2.367832 |
| 0.402856 | 0.431765 | 0.436849 | 0.417407 | 0.388274 | 0.356353 | 0.319556 | 0.279131 | 0.236783 |
| 0.5607 | 0.5482 | 0.5286 | 0.5031 | 0.4782 | 0.4560 | 0.4349 | 0.4148 | 0.3959 |
| 1.7441 | 1.9932 | 2.2424 | 2.4915 | 2.7407 | 2.9898 | 3.2390 | 3.4881 | 3.7373 |
| 3.59 | 3.77 | 4.11 | 469 | 5.42 | 6.29 | 7.39 | 8.84 | 10.79 |
| 5.607 | 5.482 | 5.286 | 5.031 | 4.782 | 4.560 | 4.349 | 4.148 | 3.959 |
| 0.0717 | 0.0735 | 0.0777 | 0.0848 | 0.0927 | 0.1011 | 0.1096 | 0.1180 | 0.1264 |
| 8.0239 | 7.5901 | 7.2226 | 6.9079 | 6.6429 | 6.4109 | 6.2211 | 6.0669 | 5.9411 |
| 700 | 800 | 900 | 1000 | 1100 | 1200 | 1300 | 1400 | 1500 |

for the discussion that follows. The optimum size is 900 kW for oil-fired engines, and 1000 kW for gas-fired engines. However, much larger savings are attained in the latter case. It should be noted that payback rises steeply beyond this savings maximum, as expected.

For fuel oil fired engines[5], the 900 kW engine appears to be the better solution: Net savings from cogeneration indeed reach a maximum for engine size = 900 kW. Any engine larger than 1100 kW is wasteful. At all times, the electrical energy requirements are fully covered up to 1100 kW (without excess availability), but engine fuel cost is higher for the larger engine $683,250/Y as compared to $599,023/Y for the 900 kW engine. On the other hand, net savings from cogeneration for the 1100 kW engine is $406,145, while this quantity peaks at $437,773 for the 900 kW engine. Note, that although the net boiler fuel cost decreases with

## Table 2-3. Cogeneration Analysis: Summary
## (Fuel = Natural Gas)

| EngineSize: kW | X | 100 | 200 | 300 | 400 | 500 | 600 |
|---|---|---|---|---|---|---|---|
| EngFuel,MBtu | | 12080 | 24160 | 36240 | 48320 | 60400 | 72480 |
| Ht.avail,MBtu | | 3066 | 6132 | 9198 | 12264 | 15330 | 18396 |
| Max Cogen kWh | | 876000 | 1752000 | 2628000 | 3504000 | 4380000 | 5256000 |
| kWhdisplCog | | 876000 | 1752000 | 2628000 | 3504000 | 4380000 | 5256000 |
| % kWhdispl | | 12.13 | 24.26 | 36.39 | 48.53 | 60.66 | 72.79 |
| $EngFuel(G) | A | 42280 | 84560 | 126840 | 169121 | 211401 | 253681 |
| $NetBoil(FO/LPG) | B | 863765 | 799131 | 734498 | 669864 | 605231 | 541307 |
| $TotCostFuel | | 906044.9 | 883691.5 | 861338.1 | 838984.8 | 816631.4 | 794987.8 |
| $NetSavCogen | C | 84616 | 169232 | 253847 | 338463 | 423079 | 506985 |
| 10~CCI | a | 0.933903 | 1.915053 | 2.947128 | 4.0342 | 5.180783 | 6.377268 |
| CCI-Avg | | 0.09339 | 0.191505 | 0.294713 | 0.40342 | 0.518078 | 0.637727 |
| FUR—Avg | | 0.6102 | 0.6102 | 0.6102 | 0.6102 | 0.6102 | 0.6095 |
| TLM —Avg | c | 0.2492 | 0.4983 | 0.7475 | 0.9966 | 1.2458 | 1.4949 |
| Payback | d | 2.36 | 2.36 | 2.36 | 2.36 | 2.36 | 2.37 |
| 10*FUR | b | 6.102 | 6.102 | 6.102 | 6.102 | 6.102 | 6.095 |
| $EngFu/kWh | | 0.0483 | 0.0483 | 0.0483 | 0.0483 | 0.0483 | 0.0483 |
| $TotFuel/kWh | | 1.0343 | 0.5044 | 0.3278 | 0.2394 | 0.1864 | 0.1513 |
| EngineSize: kW | | 100 | 200 | 300 | 400 | 500 | 600 |

engine size, there is increased thermal dumping for larger engines.

The total avoided cost of electricity is what contributes to the cost effectiveness of cogeneration. As mentioned, the 900 kW engine has the highest net savings figure in the range 100 kW to 1500 kW covered by this analysis. A closer analysis of the performance of cogeneration systems may be achieved by means of suitably defined parameters as discussed in the next sections.

### Performance Indexes

The following **Performance Indexes** are defined:

1.    Cogeneration (Avoided) Cost Index (CCI):

| 700 | 800 | 900 | 1000 | 1100 | 1200 | 1300 | 1400 | 1500 |
|---|---|---|---|---|---|---|---|---|
| 84560 | 96640 | 108720 | 120800 | 132880 | 144960 | 157041 | 169121 | 181201 |
| 21462 | 24528 | 27594 | 30660 | 33726 | 36792 | 39858 | 42924 | 45990 |
| 6132000 | 7008000 | 7884000 | 8760000 | 9636000 | 10512000 | 11388000 | 12264000 | 13140000 |
| 6062400 | 6762000 | 7196400 | 7323600 | 7369200 | 7369200 | 7369200 | 7369200 | 7369200 |
| 83.62 | 92.69 | 98.08 | 99.49 | 100.00 | 100.00 | 100.00 | 100.00 | 100.00 |
| 295961 | 338241 | 380521 | 422801 | 465082 | 507362 | 549642 | 591922 | 634202 |
| 487200 | 437094 | 389558 | 346705 | 305174 | 272768 | 245766 | 219204 | 193541 |
| 783161 | 775334.7 | 770079.4 | 769506.7 | 770255.1 | 780129.9 | 795407.8 | 811126.3 | 827742.7 |
| 576127 | 633678 | 669809 | 679423 | 681915 | 672040 | 656762 | 641044 | 624428 |
| 7.356437 | 8.172966 | 8.69792 | 8.829326 | 8.853109 | 8.614467 | 8.256927 | 7.903135 | 7.543739 |
| 0.735644 | 0.817297 | 0.869792 | 0.882933 | 0.885311 | 0.861447 | 0.825693 | 0.790313 | 0.754374 |
| 0.5984 | 0.5836 | 0.5623 | 0.5338 | 0.5078 | 0.4806 | 0.4554 | 0.4337 | 0.4145 |
| 1.7441 | 1.9932 | 2.2424 | 2.4915 | 2.7407 | 2.9898 | 3.2390 | 3.4881 | 3.7373 |
| 2.43 | 2.52 | 2.69 | 2.94 | 3.23 | 3.57 | 3.96 | 4.37 | 4.80 |
| 5.984 | 5836 | 5.623 | 5.338 | 5.078 | 4.806 | 4.554 | 4.337 | 4.145 |
| 0.0488 | 0.0500 | 0.0529 | 0.0577 | 0.0631 | 0.0688 | 0.0746 | 0.0803 | 0.0861 |
| 0.1292 | 0.1147 | 0.1070 | 0.1051 | 0.1045 | 0.1059 | 0.1079 | 0.1101 | 01123 |
| 700 | 800 | 900 | 1000 | 1100 | 1200 | 1300 | 1400 | 1500 |

**CCI = {\$NetSavCogen/\$TotCost}**

$ Net Sav Cogen = $ Displ kWh + $ Avoided Fuel
$ Sav Cogen = Avoided fuel cost (displaced thermal load)
$ Tot Cost = $ Cogen Fuel + $ Net Boiler Fuel

Because $Net Sav Cogen effectively defines the net cost of avoided energy, CCI increases with savings resulting from the cogeneration process.

2.   Fuel Utilization Ratio (FUR):

**FUR = {Btu eq. kWh displaced +
Avoided fuel}/(Engine Fuel)**

By definition, FUR < < 1 due to losses in the cogeneration process. It measures how effectively the engine fuel is used to generate electricity while satisfying part of the existing thermal load. Because FUR is directly dependent on the system's ability to replace purchased energy, engines designed to meet demand should have high FUR values.

3.    Thermal Load Matching Factor (TLM):

**TLM = 1 - {(Boiler fuel - Avoided Fuel)/(Boiler fuel)}**

This index indicates how well the cogeneration thermal output matches existing thermal load. Boiler fuel refers to fuel consumed by a boiler used to satisfy current thermal loads.

(1)  When TLM = 1: Avoided fuel = Boiler fuel, a perfect match.

(2)  When TLM < 1: Avoided fuel < Boiler fuel, with less than ideal thermal match.

(3)  When TLM > > 1: Avoided fuel > Boiler fuel, thermal dumping.

Thus, engines designed to meet peak demand will meet the condition that TLM > > 1, with resulting dumping of excess cogeneration thermal output. The term **avoided fuel** applies to the portion of the cogeneration thermal output which effectively meets current thermal load on the premises.

**Parametric Analysis of Cogeneration**

The preceding discussion indicates that the selection of engine size for the cogeneration system will depend on the scope of cogeneration. As is generally known, unless the thermal load matches the useful thermal output of the engine designed to satisfy the electrical demand requirements, there usually will be occasional shortages of cogenerated thermal output.

However, since cogeneration is primarily designed to displace kWhs purchased from Utility, the main objective in sizing the engine is to select a capacity that will satisfy the energy requirements, leaving the Utility the role of peak shaving in instances of peak demand or where demand exceeds the capacity of the cogenerating engine.

The performance indexes—**CCI, FUR** and **TLM**—are useful to measure the extent to which the cogeneration criteria are met and to guide the designer in the selection of a suitable engine capacity. Thus, CCI measures how well cogeneration contributes to reduce the cost of purchased energy. The index FUR indicates how effectively the cogenerating unit satisfies its double function of displacing utility-purchased kWh and of providing a suitable thermal output to meet existing thermal loads.

On the other hand, the index TLM would indicate how closely the engine thermal output matches existing thermal loads. As defined, CCI gauges the net amount of avoided cost through cogeneration. Thus, it should be able to measure the cost effectiveness of cogeneration more accurately than simple payback. Payback is simply an indication of how fast the initial investment is recovered: it is an accounting entity and it is rather insensitive to short term load variations.

The indexes defined above may be used to analyze the performance of the cogeneration engine on a periodic basis, as shown in Table 2-2 and 2-3. These tables show the various parameters CCI, FUR and TLM vary with engine size. The tables were constructed with data derived from the analysis of individual spreadsheets for each engine size. They include results for both the fuel alternatives (fuel oil and natural gas) discussed previously. The indexes listed on Tables 2-2 and 2-3 are the average of monthly figures for a 12-month period for each engine size used. They are also displayed in graphic form in Figures 2-1 and 2-2.

At all times, the index FUR for natural gas cogeneration exceeds that for fuel-oil fired engines indicating better fuel utilization for gas-fired engines[6]. As expected, the value of TLM increases continuously with engine size: however, beyond 400 kW, TLM > 1 thus indicating considerable thermal dumping.

Sometimes, thermal dumping of excess thermal cogeneration output is inevitable as long as the cost of purchased energy is reduced. For this situation, the index CCI becomes rather useful. In both fuel types, the value of CCI attains a **maximum** around 850 kW. However, the values of CCI are substantially larger for gas-fired engines than for oil-fired engines as would be expected, since net savings from cogeneration are higher in the former case (gas).

It is interesting to note that payback (PB) begins to increase as the fuel utilization ratio (FUR) drops for increasing engine size. The deflection occurs as the value of CCI approaches a maximum (beyond ca. 700 kW).

## Figure 2-1. Parametric Analysis, Oil Cogeneration

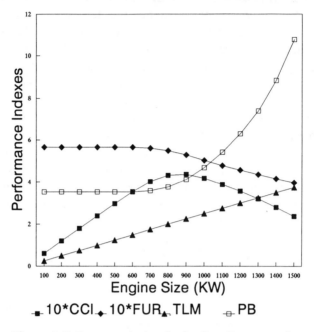

## Figure 2-2. Parametric Analysis, Gas Cogeneration

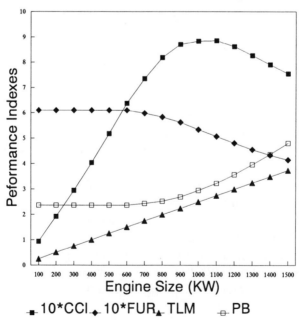

## Environmental Considerations

The main purpose of cogeneration is to reduce the cost of pur-chased energy. This objective is further reinforced by the fact that part of the existing thermal load can be replaced by the heat recovered in the cogeneration process. Additionally, from the utility view point, expecta-tions arising from demand-side management programs (DSM) have until recently led some utilities to provide incentives for cogeneration projects. However, a few utilities are now facing the prospect of excess generating capacity, thus reducing or in some case nullifying the appeal of the cost avoidance theme which has so far propelled the utilities to promulgate and promote their DSM programs.

Aside from these and related economic problems, cogeneration may contribute to an ecological problem because of deterioration of the environment by emissions from cogeneration systems. This potential for pollution would become particularly significant when poorly managed cogeneration systems operate under far from ideal conditions thus con-tributing to the contamination of the environment with toxic effluents.

As is well known, the combustion of fossil fuels when properly done yields $CO_2$ and, in the case of hydrocarbon containing fuels, water vapor as well. Thus, even under ideal conditions, burning fossil fuels always raises the $CO_2$ content in the atmosphere, with an increase in the 'greenhouse' effect. There is also a direct thermal pollution effect due to the heat carried by the hot exhaust gases. Under worse conditions, sul-fur-containing fuels may add sulfur dioxide to the environment as an air pollutant, which when deposited on the soil increases the acidity of the medium, thus altering the pH balance of the soil and contributing to the corrosion of pH sensitive materials.

One of the authors[7] has examined the inherent contamination po-tential of the ideal combustion of fossil fuels. Based on this study, a **$CO_2$ emission index** is defined as follows:

$$\delta = 1000 * (3.664 * c)/HHV^8$$

where:

      c = Carbon content in the fuel, lbmC/lbmFuel
  HHV = Higher heating value, Btu/lbmFuel

Values of $\delta$ have been calculated for various types of fuel. Based on this Emission Index, we can classify the various types of fuel under three broad categories:

| Fuel Category | Type | Emission Index[9] |
|---|---|---|
| Low emission | Molecular fuels | $\delta < 0.160$ |
| Med emission | Fuel Oils | $0.16 < \delta < 0.180$ |
| High emission | Coals | $\delta > 0.180$ |

As expected, this tabulation suggests that $CO_2$ emission in a given fuel increases with its carbon content. Thus, coals would be the worst offenders. Therefore, the index $\delta$ may be used as a measure of the potential environmental effect of any given fuel.

The assumption of ideal combustion conditions is unattainable under practical operating conditions of any engine, including those used as prime-movers in cogeneration systems. Without embarking into a complex environmental analysis of the kinetics of fossil fuel combustion, one can restrict the analysis to two major contaminants that would arise directly from incomplete combustion of fossil fuels:

Formation of carbon monoxide: $C + 1/2\ O_2 - > CO$

Water gas effect: $H_2 + CO_2 - > CO + H_2O$

The water gas effect is more pronounced in natural gas because of its higher hydrogen content relative to its molecular weight as compared to other hydrocarbon-based fuels. The phenomenon of incomplete combustion is rather complicated and is beyond the scope of this paper. A suggestion is made here that one of the reasons for the incomplete combustion of methane may be ascribed to the higher diffusivity of both $CH_4$ and hydrogen relative to that of the other molecules (such as the components of air) in the combustion mixtures. Hence, these two molecules tend to diffuse ahead of oxygen in such mixtures, thus favoring a secondary reaction with the combustion products, viz. $CO_2$, to yield among other products, water gas.

Therefore, attention must be paid to the proper operation of the cogeneration engine, particularly in the case of gas-fired engines in order to reduce adverse environmental effects due to the cogeneration process. Of course, the presence of sulfur would tend to aggravate the situation by the introduction of a highly corrosive component to the combustion effluent gases.

Because cogeneration systems operate at relatively high temperatures, other noxious pollutants, such as $NO_x$ and organic peroxides and hydroperoxides, are produced which will add to further contamination of the environment. This fact is taken into consideration in the ranking system used by NYSEO in the ICP Program which penalizes those cogeneration proposals that present the prospects of a higher environmental impact.

The other environmental aspect, which is easily overlooked, is that of thermal pollution which stems from the exhaust heat carried into the atmosphere by the effluent gases. To reduce this effect, properly designed heat sinks should be included in the cogeneration system at the time of its installation. A beneficial effect of such heat sinks would be the partial recovery of low-grade heat which can be used to preheat combustion air or other process fluids. The combined effect of cogeneration heat recovery with the heat sinks can significantly reduce the environmental impact of the process of cogeneration in comparison with the power generating systems it is intended to displace.

## SUMMARY AND CONCLUSIONS

The analysis presented in the preceding sections suggests that there are certain performance parameters that may be defined from data calculated by the Cogeneration Template that are very useful in evaluating the performance of the various alternatives to a given Cogeneration Scenario. These are:

CCI  =  cogeneration cost index =
     =  {$ Net Sav Cogen/$Tot Cost}
FUR  =  fuel utilization ratio =
     =  (Btu equiv. of kWh displaced + Avoided
        fuel}/(Engine Fuel)
TLM  =  thermal load match =
     =  1 − {(Boiler fuel − Avoided Fuel)/(Boiler fuel)}

In particular, the variations in CCI values for various engine sizes yield better indication of the performance of the different alternatives, particularly when coupled with an analysis of the other parameters, FUR and TLM. Altogether, the analysis based on such performance indexes provides a more solid basis for the selecting of the most cost effective cogeneration system.

For the illustration used in this paper, application of this methodology leads to selection of the appropriate engines for either fuel-oil-fired or gas-fired cogeneration engines. Selected data are shown below:

COGENERATION ALTERNATIVES

**Alternative 1**: Fuel = Natural Gas

ENGINE SIZE

|            | 950 kW    | 900 kW    | 850 kW    |
|------------|-----------|-----------|-----------|
| kWh/Y      | 7274400   | 7196400   | 7044000   |
| NetSavings | $675,639  | $669,809  | $656,576  |
| Boiler fuel| $368,132  | $389,558  | $413,099  |
| Payback    | 2.81      | 2.69      | 2.59      |
| FUR        | 0.548     | 0.562     | 0.575     |

**Alternative 2**: Fuel = #4 Oil

ENGINE SIZE

|            | 950kW     | 900 kW    | 850 kW    |
|------------|-----------|-----------|-----------|
| kWh/Y      | 7274400   | 7196400   | 7044000   |
| NetSavings | $433,255  | $437,773  | $437,003  |
| Boiler fuel| $422,097  | $443,092  | $464,087  |
| Payback    | 2.81      | 2.69      | 2.59      |
| FUR        | 0.548     | 0.562     | 0.575     |

As expected, Alternative 1 is the most cost effective. This table also indicates that net savings from cogeneration peak around 900 kW with a payback around 3.5 years (4.11 years for oil and 2.7 years for gas). Alternative 2 was included in the event of unavailability of natural gas. This situation actually occurs for the complex building for which the current analysis is a simplified adaptation. Because two types of fuels were used for thermal loading, the analysis was carried out on a boiler fuel cost which is a composite of the unit costs for fuel oil #4 and LPG. The economic feasibility of this alternative predicates upon the possibility of buying the fuel at a bulk price, ideally lower than the current lot price.

Since attaining maximum savings is the economic criterion, it is concluded that the smaller engine for either type of cogeneration fuel that would yield the highest savings (about 900 kW) would appear to be the most cost effective. Furthermore, the 900 kW value is the threshold beyond which payback rises rapidly. The analysis presented in this paper suggests that payback is not very effective in the evaluation of cogeneration systems. Certainly it is not to be used as the sole criterion for selecting cogeneration engine size.

The methodology used so far in the analysis of cogeneration systems may soon have to be revised to include savings resulting from demand reduction, as per mandate from DOE as expressed in Document 10 CFR Part 455 (Docket No. CE-RM-91-130) issued by the Office of Conservation and Renewable Energy, US Department of Energy. Conformance with the new guidelines may foster the need for an extensive overhaul of the SEO technical review process used in the evaluation of cogeneration applications submitted to the ICP Program which is supported by federal funds.

For instance, demand savings and cost avoidance projects may henceforth be favored even though their cost effectiveness as expressed by the simple payback criteria may be deemed marginal under current policy. New ICP Regulations and Procedures are under study by New York State Energy Office but as of this date it is not possible to predict how deeply the criteria for evaluation of cogeneration will be affected by the forthcoming changes.

Concerns with environmental contamination stemming from the combustion of fuels used in powering cogeneration system brings further emphasis on the proper management of the operation of such systems. Thus environmental considerations will reinforce the need for proper handling of the cogeneration process, beyond those of purely economic nature (such as cost avoidance and DSM incentives). However, concerns for environmental purity may not necessarily increase the cost of cogeneration, since an improvement in combustion conditions and the adoption of heat sinks (waste heat recovery) may contribute to an increase in overall efficiency of the process.

# ACKNOWLEDGMENTS

The authors wish to thank the NYSEO management for allowing publication of this Paper. The Cogeneration Template is the result of the joint collaboration of both Authors. The Parametric Analysis concept was developed as part of private research carried out by the senior author (CEG) and was initially applied to the evaluation of a cogeneration system for a large complex building situated in one of the all-year round resorts in New York state. The authors also wish to acknowledge the assistance provided by Mr. Tom Eapen, also of NYSEO, who stressed the importance of environmental effects of cogeneration and provided updated information on changes in the ICP Program guidelines arising from DOE's final ruling on the ICP Grants as expressed in Document 10 CFR Part 455 (Docket No. CE-RM-91-130) issued by the Office of Conservation and Renewable Energy, USDOE.

**Notes**
1. State University of New York.
2. Proposed cogeneration capacity: MW = megawatt; kW = Kilowatt
3. NYSEO uses as cost of electricity the average cost calculated by dividing the total energy consumption by the combined billed cost of energy and demand.
4. Discussed in the Analysis section.
5. The combined fuel is the composite of the following:
   Fuel Oil #4        261,803Gal/Yr    $194,126
   Liq. Propane     135,233Gal/Yr    $116,253
   The LPG was converted to Equivalent Fuel-Oil by using the ratio of their HHV values (95457/145000=0.658).
6. Unfortunately the scale used does not indicate the differences as clearly as the values shown on Table 2.
7. C.E. Gracias, unpublished study, "$CO_2$ Emissions and the Heating Power of Fuels - A Brief Analysis," Sep/89.
8. Similar indexes may be defined for other combustion products, such as $SO_2$ (sulfur dioxide) and CO (carbon monoxide) by adjusting the parameters accordingly.
9. For those who are chemically oriented, the trend observed for the $\delta$ values can be readily explained by comparing values with the corresponding carbon:hydrogen ratio for a given fuel.

## Appendix—List of Tables and Figures

## Bibliographic References

"New York State Energy Office's Experience with Retrofit Cogeneration Under the Institutional Conservation Program," R.P. Stewart and Naresh Ghiya, Xl WEEC Proceedings, 1988, p. 359.

"A Cogeneration Template" - Status Summary, C.E. Gracias and L.S. Hoffstatter, NYSEO, Apr. 1988, "Small-Scale Cogeneration: NYSEO Seminar Notes, 1992," L. Hoffstatter.

*Guide to Natural Gas Cogeneration*, ed. N. Hay, AGA, AEE Energy Books, Lilburn, GA 30226.

"Gas Turbine Cogeneration - Design, Evaluation and Installation," Robin Mackay, The Garret Corporation, Los Angeles, Feb. 1983.

"Cogeneration: Regulation, Economics and Capacity," Resource Dynamics Corporation, McLean, VA 22101, DOE/NBB-0031, Apr. 1983.

"Making Cogen Practical for Various Commercial Buildings," G. Meckler, Cons. Spec. Eng., Jan. 1987, p.73.

"Industrial and Commercial Cogeneration Case Studies," Synergic Resources Corporation, Bala Cynwyd, PA 19004, EPRI EM-5083, Mar. 1987.

"Proceedings: 1986 EPRI Cogeneration Symposium," Synergic Resources Corporation, Bala Cynwyd, PA 19004, EPRI EM5285, Jun. 1987.

"Cogeneration And Utilities: Status and Prospects," Synergic Resources Corporation, Bala Cynwyd, PA 19004, EPRI EM6096, Nov. 1988.

"Cogeneration - Generating Power and Profits" Engine Heat Balance, Cooper-Bessemer Reciprocating, Grove City, PA 16127.

"Handbook of High-Efficiency Electric Equipment and Cogeneration System Options for Commercial Buildings," N. Richard Friedman, Resource Dynamics Corporation, and M.H. Blatt, EPRI, EPRI CU-6661, Dec. 1989.

"All Fired Up About Cogeneration," Newsfront, Chem. Eng., Jan. 1992, p. 39.

"Cogeneration - A Technology That Has Come of Age," R.L. Oliverson, Cons. Spec. Eng., Jul. 1987, Editorial.

"Cogeneration Now Feasible For Small Buildings, W. Endicott, Cons. Spec. Eng., Jul. 1987, p. 54.

"Packaged Cogeneration - A Maturing Technology," L.J. Kostrzewa and K.G. Davidson, *ASHRAE Journal*, Feb. 1988, p. 22.

"The Challenge of Packaged Cogeneration" J. Douglas, *EPRI Journal*, Sept. 1988, p. 29

"Packaged Cogeneration for Small Buildings," J.A. Orlando, HPAC, Dec. 1988, p. 49

"Requirements for Cogenerating in Parallel With The Utility," J. Daley, EC&M, Apr. 1988, p. 63.

"Small Cogeneration System Costs and Performance," Science Applications International Corporation, EPRI EM-5954, Aug. 1988.

"Cogeneration - An Up and Coming Technology," Milton Meckler, Cons. Spec. Eng., Mar. 1989, p. 82.

"An Engineer's Guide to Cogeneration," Editors, HPAC, Jul. 1990, p. 84.

"Cogeneration Plant Slashes College's Electricity Costs," ITT Bell & Cossett, HPAC, Dec. 1990, p. 21.

"The Cogeneration Connection," J. Sinclair, HPAC, Dec. 1990, p. 43.

"Wastewater Treatment Plant Benefits from Cogeneration," B.L. Steiger, HPAC, Dec. 1990, p. 55.

"Cogeneration Revisited," Letter, P.O. Marron, Marron Associates, Eng. Systems, Jan./Feb. 1992, p. 21.

Document 10 CFR Part 455 (Docket No. CE-RM-91-130) issued by the Office of Conservation and Renewable Energy, US Department of Energy.

"Btu Tax Revisions Boost Cogen/IPP Industry," A. Donnelly and A. Wetz, *World Cogeneration*, May/June 1993, p. 1.

*Chapter 3*

# A Cogeneration Feasibility Study

Chapter 3

# A Cogeneration Feasibility Study

*James R. Geers, PLM Technologies*
*Carl D. Svard, P.E., Integrated Planning & Engineering*

H ow does one structure a feasibility study for a multi-building cogeneration system? It's an interesting challenge. Certainly there are no hard-and-fast rules or patterns. The variables involved are infinite.

For example: The cost analyses in this chapter which refer to "escalating costs of electricity," made just three years ago, today need to be ammended.

The basic purpose of this study was to determine the feasibility of cogeneration located in Tallahassee, Florida. Tallahassee is the capitol of the State of Florida with a large number of state buildings downtown, referred to as the Capitol Center, and two large Universities.

The scope of the study included the following:

Cogeneration for the Capitol Center alone, as well as adding Florida State University (FSU) and Florida A&M University (FAMU) thermal loads to the plant were also studied. Both of these Universities are within 5,000-7,000 feet of the new plant site.

The consultants were to provide a general concept, recommendations and back-up analysis in a final report. This has been Phase I of what is intended to be a four-phase project. Since cogeneration is feasible, the state intends to approve a Phase II effort to determine the financing option, participants, and a final concept design.

The State of Florida procured these A-E services through Sunbelt Engineering, Inc. (Sunbelt) and assigned Mr. Robert W. Mohrfeld of the

Department of Management Services as the Project Director. Sunbelt subcontracted to PLM Technologies, Inc. (PLM) to lead the study phase while Integrated Planning & Engineering, Inc. (IPE) assisted PLM throughout the study.

## A. Statement of Services - Phase I
The following summary is included for quick overview.

1. Collected and projected thermal and electrical load data for all three locations.
2. Assessed electrical, heating, and cooling load data.
3. Performed field survey of mechanical rooms, existing chilled water and steam plants, potential plant sites, and routing of distribution lines.
4. Assessed fuel options for cogeneration plant.
5. Developed plant alternatives and selected equipment types.
6. Selected absorption and centrifugal chillers and steam turbines.
7. Developed preliminary O&M and capital budgets.
8. Performed economical analysis and power plant modelling.
9. Prepared a preliminary report.
10. Presented preliminary report results.
11. Submitted a final study report.

## B. Status of Project
The preliminary report was submitted on February 1, 1993, and presented to various state agencies and organizations on February 11. The presentation was followed by individual meetings with FSU, FAMU, and the City of Tallahassee, as well as a joint meeting with all parties during the week of March 1-5. The final report was submitted on June 1,1993.

The starting date of the Phase II effort has not been determined. In addition, the city has retained the services of two outside engineering firms to review and comment on the preliminary and final reports. The state and city also intend to procure an independent third party to review and comment on the report.

Depending upon the participants, the cogeneration plant could serve the Capitol Center only, Capitol Center and FSU or FAMU or, as recommended, all three sites. The Phase II effort will take the project through full concept design, education of all participants, development of four working sub-groups, overall concept for the cogeneration agreement with the city, and other major project tasks.

Phase II will last six to eight months. The Phase III effort would include the financing effort as well as full design of the plant and distribution systems. Phase III is estimated to take twelve months. Phase IV would be the testing and commissioning of the plant. This final phase would take two months and be completed by early 1996.

# COGENERATION PLANT ALTERNATIVES

## A. Description of Alternatives

1.  General Overview

    The selection of the central plant alternatives was complicated by the fact that the load data for the Capitol Center and FAMU was not very specific. However, the alternatives were developed using the steam and cooling base loads at each site. The new 33 MMBH boilers in the existing heating plant were included in all alternatives as backup boilers to the new plant.

    There are several potential cogeneration plant alternatives which could serve the thermal and electrical loads of the Capitol Center, FSU, and FAMU. These alternatives vary in the type and the quantity of equipment and level of thermal and electrical output.

    However, all provided a significant and adequate supply of thermal and electrical requirements at each site. Plant configurations include variations of the following types of equipment:

    a   Gas-turbine generators with heat recovery boilers.
    b.  Supplemental firing for the heat recovery boilers.
    c.  Condensing steam turbine generators.
    d.  Electric driven centrifugal chillers.
    e.  Steam driven absorption chillers.

2.  Basecase - Capitol Center - Replacing Existing Chillers In-Kind

    This case basically includes the replacement of all building chillers over a 30-year period. They were grouped into 5-year replacement

cycles based upon age and efficiency. The total chiller tonnage of
this alternative was 11,430 tons.

3.    Alternative A - Capitol Center - Central Cooling Plant Only
      This alternative was evaluated, but the annual savings of installing
      a new chilled water distribution system and chiller plant could not
      offset the large capital investment. The payback far exceeds the
      other alternatives. This concept works well when the project in-
      volves all new buildings and the distribution system can be a part
      of the central plant complex.

4.    Alternative B - Capitol Center - Central Cogeneration Plant
      There were three sub-alternatives selected and each included four
      1500-ton electric chillers and one 3750 kW gas turbine. Each sub-
      alternative also included combinations of the following equipment:
      a 1000-ton electric chiller, a 1000-ton absorption chiller, a 1200 or
      3000 kW steam turbine and supplemental firing on the gas turbine.
      This alternative is feasible.

5.    Alternative C - Capitol Center/FSU - Central Cogeneration Plant
      There were three sub-alternatives which included either four or five
      1500-ton electric chillers; one 1200-ton absorption chiller (single ef-
      fect); two or three 1200-ton absorption chillers (double effect) at
      FSU; two gas turbines with supplemental firing and two 1200 kW
      steam turbines. The maximum steam to FSU was set at 50,000 PPH
      or 75,000 PPH. This allowed the equipment to supply the maxi-
      mum thermal requirement at FSU, which dropped the payback of
      these alternatives. This alternative is also feasible.

6.    Alternative D - Capitol Center/FSU/FAMU Central Cogeneration
      Plant - Preferred Alternative.
      There were three sub-alternatives for this option, which involved
      serving all three sites. The equipment for the preferred sub-alterna-
      tive (D-0) included: four 1500-ton electric chillers, and one 1200-ton
      double effect absorption chiller at the new cogeneration plant; two
      1200-ton double effect absorption chillers at FSU; and one 1200-ton
      double effect absorption chillers at FAMU.

      There would be four 3750 kW gas turbines. The maximum steam to
      FSU plus FAMU was set at 75,000 PPH. This allowed the equip-

ment to supply the maximum thermal requirements which again reduced the economic payback. This alternative is also feasible and was selected as the preferred alternative.

B. **Alternative Considered -**
**Base Case vs. Preferred Alternative Only.**

1. General
   All alternatives for the new cogeneration plant were specified for a 30-year economic life. All replacement costs were determined based on recommended major overhauls or complete replacements.

2. Base Case: Replace Existing Chillers In-Kind
   The base case scenario assumes a "business as usual" situation. Each of the Capitol Center buildings would continue to be served by individual electric chillers and gas/oil fired boilers. The central steam system from the heating plant would continue to serve the steam/hot water requirements of eight buildings. The electrical requirements for the Capitol Center include buildings served by the state owned distribution system plus buildings served directly by the City of Tallahassee.

3. Alternative D: Capitol Center Central Cooling Plant with Cogeneration and Steam Generation for FSU and FAMU
   The selected alternative embodies the following components:

   a. *A New Central Cooling Plant.*
      The central cooling plant would contain four to five electric-drive, high-efficiency, centrifugal chillers. A new direct-buried primary/secondary chilled water distribution system would be installed to provide chilled water to all of the Capitol Center buildings, including the "distant buildings": the FSU Law School, Law Library, Center for Professional Development, and the Tallahassee Leon County Civic Center.

   b. *A Central Cogeneration Plant.*
      The Capitol Center cogeneration plant would consist of four 3,750 kW gas-fired turbine generators with waste heat boilers for the production of steam. A steam-absorption chiller would

be utilized in addition to four electrical chillers to generate chilled water for use in the Capitol Center through the new distribution system outlined above. During Phase II additional absorption chillers would be modelled for an optimum solution.

c.  *Steam Turbines and Supplemental Firing.*
    In sub-alternative D-1, one of the gas turbines is equipped with supplemental firing to help drive two 1,200 kW steam turbines, permitting all electrical chillers at Capitol Center. Life cycle analysis showed this additional equipment actually resulted in a longer payback than the preferred DO alternative.

d.  *Steam Distribution to FSU & FAMU.*
    A new steam supply and condensate return system would be installed, in alternative D, from the cogeneration plant to the central boiler/chiller plants on FSU and FAMU campuses.

    The generators in this alternative are configured to supply, from waste heat recovery alone, the predetermined load of 75,000 PPH of steam for FSU + FAMU, plus Capitol Center's heating requirements, plus one absorption chiller at the Capitol Center.

    This results in a surplus of electrical energy available to the state for use at FSU and FAMU.

## C. Selection Analysis

1.  Energy Analysis
    The various alternatives studied involved the following datasets (combinations of individual load profiles), which were used in the cogeneration modeling program called PE-CUBE as synthetic profiles.

    a.  *Electrical*
        The study assumes that the only electrical demand directly served by the cogeneration plant is that of the Capitol Complex, not including the distant buildings.

The assumption is also made that excess electrical energy generated by the plant is fed into the city electrical grid to FSU/FAMU for credit. The credit is equivalent to the applicable energy rate ONLY, with no demand component. If the city rejects this approach, the state would install electrical feeders to FSU and FAMU.

b. *Chilled Water*

The modeled chilled water load for the central plant includes the composite chilled water load from the 21 Capitol Center buildings, plus the composite cooling load from the distant building load.

The modeled chilled water loads for the FSU and FAMU central plants were converted to steam equivalents (at 12 lbs of 150 psig steam per Ton of cooling), enabling these loads to be combined with the hot water/steam loads.

c. *Hot Water/Steam*

For hot water/steam, the Capitol Center is always modeled using the buildings which are to be served by the central boiler plant. It is assumed that the other buildings in the Capitol Center will continue with their current individual heating configurations.

The HW/steam loads of FSU and FAMU were combined with the steam-equivalent chilled water loads, to give an overall steam demand for those sites. It was then possible to model a steady supply of steam that met the cooling and heating needs of all sites.

2. Equipment Sizing Philosophy

Specific equipment choices were selected for modeling in PE-CUBE.

a. *Electrical*

All electrical generation is supplied by multiples of the SOLAR Centaur "H" gas turbine, at times with supplemental firing of the waste heat resource.

b. *Hot Water/Steam Backup*
The pair of existing boilers at the central boiler plant is available in all alternatives as part of the backup to the new plant.

c. *Chilled Water*
Two types of chillers were modeled: electrical driven centrifugal, and steam driven absorption. All chillers were modeled with built-in electrical auxiliaries for chilled water distribution requirements.

3.  Cogeneration Plant Modeling and Analysis
This section summarizes the configuration for the selected "D" alternative.

a. *Alternative D-0*:
(4) Electrical Chillers + (1) Absorber at Capitol Center, 75,000 PPH Steam to FSU+FAMU.

In this alternative, four gas turbines supply electricity to meet all Capitol Center needs (including four 1500T electrical chillers) and generate waste heat steam to power one additional 1,200T steam absorption chiller at Capitol Center, and a predetermined steam flow of 50,000 PPH to FSU, and 25,000 PPH to FAMU.

# FUELS

The types of fuels available for use at the site were analyzed. Natural gas and fuel oil were considered as the primary energy source of the plant. It appears that the best fuel for use in the plant would be natural gas. It is available within the Capitol Center; it has the lowest cost, and is readily available from the city. Its clean burning characteristics reduce the maintenance and environmental permitting problems. Interruptible gas service offers additional cost savings.

No. 2 fuel oil should be considered as the backup energy source. No. 2 fuel oil is a relatively clean fuel, is readily available, and is easy to store. The greater cost of this fuel is offset by the low expected usage rate due to the demonstrated (10-year) reliability of the natural gas service in the Tallahassee area. We have assumed that a 10-day fuel oil storage supply will be installed.

# ENERGY LOADS - SUMMARY

Summaries of peak and minimum load demands for each of the energy quantities for each site are presented below in Table 3-1.

**Table 3-1.**

| Energy Demands | Capitol Center | FSU | FAMU |
|---|---|---|---|
| **Electrical Demands** | | | |
| Maximum | 13,471 kW | | |
| Minimum | 4,450 kW | | |
| **Hot Water/Steam Demand** | | | |
| Maximum | 21,700 #/hr | 72,000 #/hr | 30,000 #/hr |
| Minimum | 900 #/hr | 10,100 #/hr | 12,300 #/hr |
| **Chilled Water Demand** | | | |
| Maximum | 6,200 Tons | 4,800 Tons | 2,000 Tons |
| Minimum | 670 Tons | 1,000 Tons | 500 Tons |

# CAPITAL COST ESTIMATES

The cost estimating approach was straightforward. The estimates were developed using means cost data, historical figures, recent bids within the Capitol Center for a similar gas turbine, and using Sunbelt's knowledge of the labor and material costs for northern Florida. These cost estimates are only "study" quality.

Once a specific concept is selected then a plus or minus 15% cost estimate will be developed during Phase II. The total capital cost varied from $14,250,000 to $51,750,000. The gas turbine cost was within plus or minus 5% since Sunbelt had just bid a similar gas turbine for Baptist Medical Center in Jacksonville.

The contingency for each alternative varied between 15% to 25%. The lower percentage was used on the large capital investment while the higher percentage was used on the lower investments. The contingencies were varied during the analysis, but did not significantly increase or decrease the payback for the alternatives.

It was assumed that the state would finance this plant as a normal budget appropriation. Financing cost and interest during construction were not included.

# LIFE CYCLE COST ANALYSIS

A life cycle costing program developed by PLM was used throughout the analysis. Significant effort was expended in performing a sensitivity analysis of the key economic parameters to determine the impact on the economic viability of each alternative.

Escalation rates for the energy categories follow the average values used for these factors. For electrical energy 3%, 4%, and 5.7% escalation rates were used in order to see their effect on the payback. The conservative value of 4% was selected for the summary figures.

The payback varied from just over 3 years to more than 30 years, using the annualized method, and from 6.5 years to over 30 years using the actual cash flow method. The district cooling plant concept had the longest payback and was eliminated from further consideration.

*The cogeneration alternatives for the Capitol Center have acceptable paybacks and, if the Universities elect not to participate, then this option is still an economical choice.*

However, the FSU and FSU/FAMU alternatives all have very similar paybacks and either one or both could participate without significantly impacting the payback of the project. The shorter payback for these alternatives basically results from offsetting higher-cost electrical energy with lower-cost natural gas, and sizing the plant to supply almost all of the base energy load for the Capitol Center, FSU, and FAMU.

The state has established 15 years as the maximum payback for this type of project. All of the cogeneration alternatives meet this criteria. The state should use both the annualized and actual payback methods when it is making a decision on this project. Tables 3-2 and 3-3 below provide a summary of parameters and paybacks.

**Table 3-2. Life Cycle Cost Analysis**
*Economic Parameters Summary*

| | | |
|---|---|---|
| Yearly Discount Rate (%) | | 8.2% |
| Number of Alternatives | | |
| Analyzed (including base cases): | 15 | |
| Economic Life (years): | 30 | |

| Energy Source | Current Cost ($/MMBtu) | Yearly Rate |
|---|---|---|
| Electricity | $20.22 | 4.00% |
| EL Credit | $16.69 | 4.00% |
| Natural Gas | $2.75 | 3.00% |

**Table 3-3. Life Cycle Cost Analysis**
*Summary of Payback - Preferred Alternative*

| Payback Options | EL% | D-0 |
|---|---|---|
| Annualized | 3.0 | 5.9 |
| | 4.0 | 4.6 |
| | 5.7 | 3.1 |
| Cash Flow (Actual) | 3.0 | 7.7 |
| | 4.0 | 6.9 |
| | 5.7 | 6.5 |
| Special Cases (all on annualized basis) | | |
| No Demand Side Rebate | — | 4.8 |
| No EL Credit for excess back to utility | — | 8.9 |
| Worst Case Scenario | — | 13.4 |

# DISTRIBUTION SYSTEMS

The distribution systems evaluated in this study are a high capital cost. However, if designed and installed by a qualified contractor, these systems will provide a long life. Proper maintenance is also required.

The central heating plant distribution system located within the Capitol Center will continue to be a part of all alternatives. The plant will provide backup steam service whenever the new plant needs the steam for base load or under planned maintenance or emergency shutdowns. This system was not expanded for purposes of this study.

All of the cogeneration alternatives include a new direct buried chilled water distribution system for the Capitol Center. The direct buried approach is recommended and will reduce the initial capital cost. All of the mechanical rooms in the Capitol Center buildings would be modified for a centralized distribution system. These costs were included in the life cycle costing analysis.

The alternatives which included cogeneration also included a new lift-top tunnel system from the plant site locations to the main steam line outside the heating plant. This connection will allow the new plant to pick up the steam load which was being served by the new natural gas-fired boilers.

The alternatives which include FSU and FAMU have new steam lines in lift-top tunnels from the plant site to the central plant at each campus. These lines will provide the base steam load at each site and additional steam which will be used for the production of chilled water via new absorption chillers at each campus. The new tunnels will be similar to the tunnels recently installed in the Capitol Center.

# SITING

Although there were several sites considered during the initial stages of the study only four remain based on the evaluation to date. The effort to date can be classified as preliminary at best. Three of the locations are on the extreme south boundary of the area referred to as the Capitol Center while the fourth is on the property just east of the city's substation number 6.

There may be other locations, but the limitations of this study restricted the effort to basic evaluations. Some proposed sites were eliminated due to soil contamination while others were within restricted areas based on the recently published state/city master plan. A hard siting effort will occur during the initial stages of Phase II. The basic building varies from 17,000 square feet to 35,000 square feet, depending upon the alternative.

# PURCHASED UTILITIES

Currently the city does not have any cogeneration plants connected to its system. Therefore, extended negotiations will be necessary with the city. The city is actively pursuing new baseload generation capacity for the 1996 thru 2010 time frame. This new plant could be on-line by 1996 if the decision to proceed were given during fourth quarter of 1993.

The new cogeneration plant will significantly change the mixture of energy purchases from the city. The plant will be a base load plant which will use large quantities of natural gas but minimum amounts of electrical energy will be purchased for the Capitol Center.

In addition, most of the alternatives produce excess electrical energy which can be used to offset purchases at FSU and FAMU. The rates for the energy produced at the plant will be determined as part of the Phase II effort. There are several methodologies which can be used.

Assuming that the state owns/operates this plant, new electric, natural gas, and cogeneration agreements must be negotiated with the city. If the state elects to have the city own and operate the new plant, then the savings will be lower because the state would not have to operate this plant. The economical, political and other factors needed to make this decision must be carefully considered by the state.

Presently, the city's position is unknown. They have evaluated the plant from a "utility perception" and found it uneconomical. However, they are currently evaluating the economics based on a state/city partnership.

# OPERATIONS AND MAINTENANCE

**General**

Cogeneration plants require a qualified staff and an annual operating budget to operate and maintain the equipment and building. The magnitude of each will vary depending on the type and size of the plant. Considering the size of this proposed plant and the facilities which it will serve, it is assumed that an unmanned plant is not desirable due to the importance of steam and chilled water throughout the sites, and the need for a rapid response in case of failure of any system.

Only continuously manned plants have been considered. The total O&M Budget for the district cooling/steam plant referred to as alternative A is approximately $600,000 per year. The preferred alternative budget is $1,500,000.

It is assumed that major repairs will be accomplished by an annual maintenance contract, or by contracting out the work on an as needed basis. The recommended staff was for operation of the plant on a daily basis and performance of the necessary routine maintenance.

### O&M Budget

The operating and maintenance costs have been estimated based on a similar plant in the Jacksonville, Florida, area. Staff salaries and benefits were estimated along with the maintenance costs.

### Service Contract (Gas Turbine Alternatives Only)

The utilization of a service contract arrangement in connection with the new cogeneration plant is recommended if the state operates the plant. A 5-year service contract with a 5-year option (if possible) will provide the ability for the state to ensure a reliable operation and provide time to train and license their own operators.

If the training of in-house personnel does not work out, the state can exercise the 5-year option and ensure continued operation. The intent was to provide for the safe operation of a multiple gas turbine cogeneration plant which must be staffed with highly qualified personnel.

# ENVIRONMENTAL PERMITTING

### Overview

The following discussion addresses the types of environmental permits required and provides some insight into the specifics of what may be required, based upon conversations with regulatory agency representatives. This discussion is presented in two subsections: analysis of permitting requirements and anticipated permit schedule.

## Analysis of Permitting Requirements
*General*

Agencies that could be involved in the permitting process, include the Public Service Commission (PSC), Florida Department of Environmental Regulation (FDER), Florida Department of Transportation (FDOT), and local county/city permitting agencies, including county/city road departments (where off-site utility construction in public rights-of-way is undertaken). PSC involvement would be restricted to the provision of electric and gas service to the new cogeneration plant. The PSC would also be involved should the state go into a program of cogeneration and make excess power available beyond its own uses.

FDOT involvement would occur for any gas transmission facilities. However, if the supplier of the gas is responsible for installation of the gas transmission lines to the site, then they would assume all responsibility for appropriate permits and regulatory requirements. Similarly, work within FDOT, county or city rights-of-way will require road use/utility permits and would apply to off-site work for piping distribution systems.

*Plant Siting*

Whatever site is selected, it will trigger appropriate water and sewer main connections, relocations and/or removals as needed. It will be very important to conduct adequate site planning and field verification of all utility sizes, locations and depths early in the design to minimize the necessity of relocations, and to plan for maintaining utility services during construction. Plant siting will also require review by a number of governmental agencies, including the FDER, U.S. Environmental Protection Agency, and the City of Tallahassee.

## Anticipated Permit Schedule
*Timetables*

The timetables for preparation of the permit application forms and acquiring the approved permit vary substantially from a matter of weeks to an estimated six (6) months or more for each permit.

*Summary*

In summary, the anticipated permit schedule is estimated at 12 months for the new plant. As pointed out in the discussion, most, if not all, of the 12 months would be occupied by normal design and construction activities. For the coming changes in the permitting process do not appear to significantly alter the time factors discussed in the above.

# EXECUTIVE SUMMARY

**Summary**
*Observations*

1.  The concept of having a single cogeneration or energy plant serving multiple complexes is not new.

2.  The state should maintain the option of owning and operating this plant regardless of whether the city or a third party would be the owner/operator of the plant.

3.  The various state agencies, as well as FSU and FAMU, have significant "joint venture" issues to resolve once a decision to proceed has been made.

4.  The plant can be a showcase/training facility for both elementary, middle school, and senior high school, as well as for the FSU/FAMU engineering school.

5.  The quality of the load data at each facility was not as good as was desired, but adequate for a "feasibility" study effort.

6.  FSU and FAMU have a base steam load which can be served by the new plant. They also have a base cooling load which could be served by new absorption chillers located on each campus.

7.  The city is seeking electrical generation capacity during the late 1990's and early 2000's. A portion of that capacity could come from this new plant.

8.  A major issue of the state concerns the involvement of the city. Will the city choose to participate as owner/operator, operator, or not at all?

# CONCLUSIONS

1. Cogeneration is feasible for the Capitol Center with or without the FSU and/or FAMU loads.

2. The new plant should serve all three facility "base" loads. However, the total load at each site should be factored into the cogeneration plant service to optimize the economics of the project.

3. The majority of the cooling loads for all sites occurs during a 10 hour weekday period (50 hours per week), not a 24 hour weekday period or throughout the weekend.

4. As the result of a variable cooling load and the high cost of the Capitol Center chilled water distribution system, a new cooling-only Capitol Center central plant is not economical. However, a new chilled water distribution system can be economically justified when combined with the cogeneration plant concept.

5. The new cogeneration plant steam distribution system would tie into the existing Capitol Center steam system.

6. The Capitol Center steam distribution system would not be extended under the cogeneration project.

7. Thermal energy to FSU and FAMU campuses should be supplied in the form of steam. Chilled water production would occur at each site via absorption chillers in the respective plants.

8. Electrical energy generation should be maximized by adding thermal loads, due to the relatively high cost of electrical energy purchases. The significant operating cost savings from the new cogeneration plant result from the use of the cogeneration concept plus utilizing natural gas at \$2.75/MMBtu versus electrical energy at over \$20.22/MMBtu.

9. The concepts in this report minimize the off-site electrical purchases and maximize the on-site generation by picking up the

maximum of thermal load at the three locations. There is excess electrical energy in most all of the alternatives.

10. The general siting constraints restrict the plant sites to the southern portion of the Capitol Center area, but outside the main campus area.

11. A new steam distribution system should be installed to FSU and FAMU using a lift-top tunnel concept.

12. The chilled water distribution system should be direct buried and not be placed in common steam tunnels.

13. The Tallahassee Civic Center load is very minimal except for the large events which occur several times per month. The new plant can serve this load, but it was not a major factor in sizing the plant since it is an off-peak load for most workdays.

14. The new plant should be developed with multiple units in lieu of a single large gas turbine which would present serious reliability and redundancy problems when considering the critical nature of the Capitol Center loads.

## RECOMMENDATIONS

1. The state should proceed with all phases of this project.

2. The state should determine the city's plans for participation as soon as possible.

3. The Universities should participate.

4. A task force should be developed with teams for policy, technical, marketing, and utility contracts. A project director should be appointed with a single state agency taking the lead.

5. The state should begin negotiations with the city so that a general cogeneration agreement can be developed.

6. The Public Service Commission should be contacted and a brief overview of the project provided. This agency will be involved throughout the project.

## Recommended Alternative

The bottom line of this feasibility study is that in all basic cogeneration alternatives, the paybacks are much lower than the maximum allowable paybacks of 15 years. Combining the Capitol Center with FSU and/or FAMU provides the paybacks. These paybacks result from the significant reduction in electrical purchases and from maximizing output on the gas turbines.

There is a significant difference in producing steam and electricity using a cogeneration process with $2.75/MMBtu fuel vs. using very costly electrical energy at $20.22/MMBtu. This difference provides the very basis of the large annual savings, and its magnitude is the reason the other cost variables have a secondary impact on the viability of the project.

The study shows that the new cogeneration plant should serve all three campuses and thus maximize the savings. Allocation of these savings will need to be determined; but, in any case, all parties will receive tremendous operational savings. The specific plant equipment selection requires further refinement which will be accomplished during the preliminary concept design effort.

The rationale for selecting the preferred alternative (alternative D) included the following

a. Provides "optimal" choice for the state.
b. Lowest overall life cycle cost with reliable operation.
c. Lowest payback, whether measured by annualized or actual methods.
d. Provides for easy maintenance of major components.
e. Provides generating capacity to the city.
f. Meets "demand-side management" objectives of the city.
g. Meets the State Energy Executive Order.
h. Avoids the replacement of individual building chillers in-kind.
i. Meets the intent of the state/city joint agreement for a cooperative effort in cogeneration projects.
j. Increases energy efficiency.
k. Enhances state control over energy prices.
l. Provides major benefits to the state and city.

**Obstacles**

The major obstacles in implementing a partnership in cogeneration include:

- Obtaining the support of the city commission/FSU/FAMU.
- Coordinating facility planning.
- Negotiating with the city.
- Securing the necessary permits and energy agreements with the regulatory agencies.
- Appointing a full-time project manager.

*Chapter 4*

# Cogeneration System Integration and Management

# Chapter 4

# Cogeneration System Integration and Management

*Dr. Robert V. Peltier, P.E., CCP*
*James F. Ring, Intellicon Incorporated*

How many times have you heard of a cogeneration plant that seems to meet its technical performance specifications yet is only a marginal economic performer? How many more systems began with outstanding economics but over a period of time increased regulatory and utility pressures made the original system design basis obsolete. A cogeneration control system designed to continuously optimize plant operating economics is without a doubt the key to economic survival in this era of dynamic rate design.

This chapter is meant to communicate to prospective purchasers or current owners of cogeneration plants, system designers and contractors several considerations for cogeneration control systems. The chapter is not written around or recommends any particular equipment and is not a specification but is meant to present current and future design issues that should be considered in the design of a modern cogeneration control system.

These design considerations are based on the combined experience of the authors that extends over all phases of the cogeneration industry and visits the authors have made to a large number and a variety of types of cogeneration plants. It is our observation that the design considerations in this chapter would benefit a large number of operating cogeneration plants. The proposed design and economic optimization considerations are being successfully used in many cogeneration plants across the United States.

The chapter is presented in three sections. The first section discusses the need for optimizing the operation of a cogeneration plant to maximize the economic return on investment. The second section describes ten design issues that should be considered in any cogeneration plant design to increase plant reliability. The final section gives an overview of the future direction of economically optimizing groups of cogeneration plants with the serving utility.

# SYSTEM OPERATION

Operators of cogeneration systems face an often complex problem when considering operating strategies for their facilities. The economic optimization of plant performance is dependent on many load and performance variables and time variable utility rate structures.

The management of the cogeneration system requires constant evaluation of these load and performance variables and the relationship of these variables to utility rates. The operating strategy used must be dynamic and available on a uninterrupted basis to meet the management goal of maximizing the plant economic return.

The benefits of complete management and control system for the cogeneration system operator are two fold. Lower initial capital cost as compared with individual electrical switchgear, process controls, and data acquisition systems results. Second, and most importantly, the day to day operation of the cogeneration system is optimized to achieve maximum economic return.

Generally cogeneration control systems are described as belonging to one of the following general categories listed in descending order of complexity:

- Manual
- Semi-Automatic
- Integrated
- Optimized

**The Manual Control System** is generally a system where manual controls on each piece of equipment locally controls the plant. The manually controlled plant may not have a central control room. The generator controls may also be manual. Few plants today are designed in this fashion and do not lend themselves to be economically operated.

**The Semi-automatic Plant** is a common plant configuration. This plant integrates the equipment local control panels in a central control center. The operator sequences the plant by the individual equipment panels which contain the equipment sequencing and protection system. Balance of plant controls and instrumentation are either locally mounted and controlled or are consolidated on one or more panels in the control center. Generator controls are typically automatic. This plant design also does not allow optimizing plant economics as the operator still controls system sequencing and loads.

**The Integrated Plant** is characterized by a completely automated operation from sequencing of start-up and shutdown to control of individual systems, subsystems and instruments. The automatic plant also typically interfaces with OEM control systems for sequencing purposes. The automatic system also includes a supervisory system of data acquisition and logging for historical and billing purposes, and problem identification. Many recently installed plants have an integrated control system in one form or another.

**The Optimized Control System** encompasses all the elements of the integrated plant but with the addition of a supervisory control mode. The supervisory mode determines the most economically advantageous method of operating the plant and adjusts the control system to "optimize" plant economics under all control modes or design plant operating conditions.

This supervisory control or "Optimized mode" considers all the plant independent variables and historical operating information and projects the near-term plant loads. The Optimizer selects the most economic operating mode and adjusts the plant control system accordingly. What are these independent variables that the optimizer must consider? Table 4-1 summarizes a number of these variables.

**Table 4-1. Economic Considerations for Optimized Plant Operation**

- Utility TOU rate structure
- Plant Load Profiles and projected load profiles
- Equipment part-load performance
- Start-up and shutdown costs
- Maintenance planning
- Other O&M costs

Many utilities employ a time-of-use (TOU) rate structure for purchase and sale of electricity. Today utilities typically employ a rate structure that rewards the cogenerator for power production (sales) during the high electrical usage (peak and semi-peak) parts of the day and penalizes the cogenerator for power sales during the off-peak periods. Therefore it is not uncommon for the optimum control mode for a plant to change from baseload operation during one part of the day to electrical load following to thermal load following at night as purchase power may be less than the cogenerator cost of electrical power production.

An economically optimized plant must also consider the range and types of thermal load profiles. For example, consider the plant that requires multiple forms of thermal energy such as steam for humidification, hot water for process use and steam for an absorption chiller. The steam use priority for a fixed available steam supply must be determined on an economic basis as a function of the cost of the alternative steam supply and as a function of the projected load requirements. The projected load profiles are usually a function of time (work day and non-work day), season (ambient temperature, humidity, etc.) and plant production capacity.

The optimized operation must also consider the part-load efficiency of the major equipment used to satisfy the plant loads. For example the efficiency of a reciprocating engine drops approximately 0.07%/% reduction in load. Except for the baseload electrical operating mode, the optimized operation of a plant in the thermal load following mode, electrical load following mode or balancing the buy-sell meter mode must consider the part-load operation (calculated based on instantaneous or performance test data and not manufacturers performance curves) of the prime mover.

Less obvious is the part-load performance of other equipment, especially an absorption chiller. An absorption chiller does not function particularly well when operated off-design. A typical 400 ton unit at full load consumes approximately 17.6 lbs/ton hr while at 75% load requires 23.4 lbs/ton hr. Economic considerations may conclude an absorber should be shut down and a centrifugal chiller started on purchase or generated power during off-peak periods. Conversely we must consider the economic impact of start-up (labor costs, maintenance costs, demand charge, etc.) and the time required to start-up equipment. Equipment is shutdown for a short period of time such as a single shift.

Similarly when a prime mover trips during a peak demand period.

The increased peak demand charge for the month can exceed several thousand dollars or more. If there is a ratchet charge the impact of a single trip could impact the plant economics for the next year. The optimizer can consider the potential costs of a service interruption versus the increased demand charge. One particular facility recognized this problem and automatically trips the complete central plant when the prime mover trips. The prime mover is then resynchronized and service restored avoiding an increased demand charge. If the prime mover cannot be resynchronized the operator can manually start the chilled water backup systems.

Another consideration for optimum operation of a plant concerns maintenance planning. When is the most economically most favorable time for routine and planned maintenance activities on a plant? Consider the inlet air filters on a gas turbine plant that must be periodically changed based on increased pressure drop. A typical gas turbine in the 3000 kW size range will lose 0.5% in power output or will lose approximately 0.2% in thermal efficiency at constant load per inch of water increased inlet pressure drop. Changing the filters may require several hours of plant downtime during which results in increased operating costs or purchase power costs. Since the performance degradation of a gas turbine is a function of the inlet pressure drop there is an economically most favorable pressure drop, day and time of day to change filters. Consideration must be given to these and many similar scenarios when economically operating a plant.

Table 4-2 summarizes a number of the plant dependent variables the optimized control system establishes.

### Table 4-2. Plant Dependent Variables Determined for Economic Plant Operation

- Load balancing method (thermal, electrical, etc.)
- Load balance of multiple prime movers
- Thermal load prioritization
- Maintenance planning

The actual economic benefits on the management and control system are most easily understood by studying an example of a representative cogeneration systems operation. The following example is based on a 650 kW reciprocating engine/generator set, a 100 ton absorption chiller, and heat recovery equipment delivering heat to three separate heat loads.

The example system at present is in operation within a utility service area which uses electrical TOU rates, including peak and non-coincident peak demand charges. Natural gas is the engine fuel. The electrical load placed on the cogeneration and utility system varies from a peak of approximately 1500 kW to a low of 300 kW. The thermal load required by the customer varies from zero to approximately two million Btus per hour.

The economics of this point of operation, specifically the net avoided cost of energy created by the system is calculated by summing the value of the electrical and thermal products and then subtracting the fuel and maintenance costs. The system full load fuel input is 8,070,000 Btu/hour of natural gas. The variable component of maintenance cost is assumed to be constant at $0.0106 per kilowatt hour produced.

The varying TOU electric rates coupled with the varying electrical and thermal loads create dramatically different operating strategies during any particular time period. To illustrate the effect of economic operation we will analyze three operating cases. Each case represents actual electrical and thermal loads with the appropriate TOU rates which apply to the point in time considered.

### Case 1—Peak Demand Period

The first point analyzed occurs at 6:00 p.m., which falls within the utility electrical peak demand period in the winter months.

Loads were recorded as follows:

| | |
|---|---|
| Site electrical demand: | 1001 kW |
| Site heating load | 114,200 Btu/hour |
| Site cooling load | 52 tons/hour |

The cogeneration system was producing the following:

| | |
|---|---|
| Electrical output | 650 kW |
| Thermal output | 1,734,000 Btu/hour |

The system electrical and thermal production was allocated as:

All electrical output was used by the site with the balance of 351 kilowatts supplied by the serving utility.

The absorption chiller, supplying 52 tons, used 967,200 Btu/hour of the thermal output, with other heat loads using 114,200 Btu/hour of the thermal output.

Excess thermal output of 652,600 Btu/hour was rejected to the atmosphere.

Value of energy produced per hour of operation:

| | |
|---|---|
| 650 kW @ 50.08011/kW-hr: | $52.07/hr |
| 52 tons of chilled water produced by the absorption chiller displaces the purchase of 52 kW hr/hr based on a centrifugal chiller @ 1 kW/ton: | $ 4.17/hr |
| 114,200 Btu/hr of heat delivered, displaces 142,750 Btu/hr of fuel input to an 80% efficient boiler at a fuel cost of $3.70/MMBtu | $0.53/hr |
| **Total avoided energy cost** | **$56.77/hr** |

System operating costs per hour of operation:

| | |
|---|---|
| Fuel; 8,070,000 Btu/hr @ $2.72/MMBtu: | $21.95/hr |
| Maintenance costs, $0.0106/kW hr @ 650 kW: | $ 6.89/hr |
| **Total system operating cost** | **$28.84/hr** |
| **Net avoided cost realized:** | **$27.93/hr** |

The system operating with these loads and utility rates is most efficiently operated at 100% of its rated output, regardless of the system thermal load. This is attributed to generation costs being less than avoided electric costs, $0.0444 and $0.08011 respectively. If excess electrical energy were available, it could be sold to the utility at $0.11808/kW hr. These operating conditions do not require the use of the system optimizing ability. Operation is very simple with the primary concern being system reliability.

### Case 2—Semi-Peak Demand Period

The second point analyzed occurs at 9:00 a.m., which falls within the utility electrical semi-peak demand period in the winter months.

Loads were recorded as follows:

| | |
|---|---|
| Site electrical demand: | 1100 kW |
| Site heating load | 248,100 Btu/hour |
| Site cooling load | 46 tons/hour |

The cogeneration system was producing the following:

| | |
|---|---|
| Electrical output: | 650 kW |
| Thermal output: | 1,734,000 Btu/hr |

The system electrical and thermal production was:

All electrical output was used by the site with the balance of the 450 kW supplied by the serving utility.

The absorption chiller, supplying 46 tons, used 855,600 Btu/hr with other heat loads using 248,100 Btu/hr the thermal output.

Excess thermal output of 630,300 Btu/hr was rejected to the atmosphere.

Value of energy produced per hour of operation:

650 kW @ $0.05053/kW-hr:                              $32.85/hr

46 tons of chilled water produced by the absorption chiller displaces the purchase of 46 kW hr/hr
based on a centrifugal chiller @ 1 kW/ton:            $2.32/hr

|  |  |
|---|---|
| 248,100 Btu/hr of heat delivered, displaces 310,125 Btu/hr of fuel input to an 80% efficient boiler at a fuel cost of $3.70/MMBtu: | <u>$1.15/hr</u> |
| **Total avoided energy cost:** | **$36.32/hr** |

System operating costs per hour of operation:

|  |  |
|---|---|
| Fuel 8,070,000 Btu/hr @ $2.72/MMBtu: | $21.95/hr |
| Maintenance costs, $0.0106/kW hr @ 650 kW | <u>$ 6.89/hr</u> |
| **Total system operating cost** | **$28.84/hr** |
| **Net avoided cost realized:** | **$ 7.48/hr** |

The lower net avoided cost at this operating point is attributed to the lower value of the electrical component of the avoided cost.

At this point were are wise to consider the consequences of reduced heat and/or cooling loads and the potential of selling excess electrical output if site load were to fall below 650 kW. Generation cost remains below avoided cost of purchased electricity, $0.0444 and $0.05053/kW hr respectively, but excess electricity can be sold at $0.042/kW hr. Consequently the output of the plant is throttled to meet the highest of the site electrical or thermal demand. This strategy would avoid losing $0.0024/kW hr sold to the utility. Throttling to the higher demand of site electrical or site thermal would be practical as the value of thermal product used, on a kilowatt hour basis, would be $0.009, which is greater than the marginal loss of $0.0024 when selling excess electricity to the utility. This situation rapidly changes in the next TOU period.

### Case 3—Off-Peak Demand Period

The third point analyzed occurs at 10:45 p.m., which falls within the utility electrical off-peak demand period in the winter months.

Loads were recorded as follows.

|  |  |
|---|---|
| Site electrical demand: | 408 kW |
| Site heating load | 1,040,000 Btu/hour |
| Site cooling load | 17 tons/hour |

The cogeneration system was producing the following:

| | |
|---|---|
| Electrical output: | 508 kW |
| Thermal output: | 1,356,200 Btu/hr |

The system electrical and thermal production was:

508 kW was used by the site with the balance of the 100 kW was sold to the serving utility.

The absorption chiller, supplying 17 tons, used 316,200 Btu/hr with other heat loads using 1,040,000 Btu/hr the thermal output.

There was no excess thermal output as the system was matching the thermal output to the site thermal demand.

The engine required 6,604,000 Btu/hr of natural gas at this load.

Value of energy produced per hour of operation:

408 kW @ $0.04250/kW-hr:                                 $17.34/hr

17 tons of chilled water produced by the absorption chiller displaces the purchase of
17 kW hr/hr based on a centrifugal chiller
@ 1 kW/ton:                                             $ 0.72/hr

1,040,000 Btu/hr of heat delivered, displaces
1,300,000 Btu/hr of fuel input to an 80% efficient
boiler at a fuel cost of $3.70/MMBtu:                    $ 4.81/hr

The excess electrical output of 100 kW sold to
the utility @ $0.02747/kW hr                             $ 2.75/hr

**Total avoided energy cost**                               **$25.62/hr**

System operating costs per hour of operation:

Fuel, 6,604,000 Btu/hr @ $2.72/MMBtu:                   $17.96/hr

Maintenance costs,
$0.0106/kW hr @ 508 kW                                   $ 5.39/hr

**Total system operating cost**                            **$23.35/hr**

**Net avoided cost realized:**                             **$2.27/hr**

If we examine the incremental values of cooling and heating it can be seen that the incremental value of recovered heat delivered summed with the value of excess electricity sold to the utility is greater than the incremental cost of generating a kilowatt hour. As long as this is true, the plant is operated at an output which meets the site thermal demand.

If we operated the engine at full load with the same thermal loads it can be shown that a loss of $255 per hour would result. In a given year for this utility, there are approximately 4700 off-peak hours. Operation of the plant at full load rather than at the specified part-load conditions results in an annual loss of $11,985. The loss is magnified if we consider the seasonal aspects of the TOU rates.

How significant is this loss? If we consider the simple economics of a system simple payback of 5 years on an investment of $1,000,000 for the plant, the annual energy savings are approximately $200,000. Economic dispatch of the plant results in 6% of the projected annual energy savings. In reality, system annual energy savings can be increased from 5% to 15% by controlling the economic operation of the plant.

The example used is a simplification of the actual plant operating conditions as it considers only a single point in time, specifically a fifteen minute interval out of 18,800 such intervals in the 4700 hours of off-peak operation. Each of these intervals have unique operating profiles which require a unique operating strategy. Similar computations and strategies must be considered with minor changes in rates and/or loads during the semi-peak period. Other significant energy costs include system demand reduction by using system generated power to operate the centrifugal chiller, site operating requirements that supersede system economics (e.g. air conditioning for a computer space, etc.), auxiliary power consumption, part load equipment efficiencies (engine, absorption and centrifugal chillers), block rate natural gas pricing and standby charges.

## SYSTEM INTEGRATION

Economic dispatch of the cogeneration plant is one part of the equation to maximize system savings. An important part of the plant economics is how reliable will the system operate throughout this and future years. Based on our design and operation experience we have identified the following ten specific suggestions that will improve system reliability and performance.

1. The key performance parameters required to do a heat and mass balance for the complete plant and all significant pieces of equipment in the plant should be data logged. For example fuel flow, A/F ratio or $O_2$ in stack, inlet air temperature, jacket water flow and inlet and outlet temperatures (inlet temperature and steam pressure for an ebullient engine), cooling water flow and temperature rise, exhaust gas temperature and electrical power produced are the minimum data for a reciprocating engine heat and mass balance. An absorption chiller heal and mass balance requires the chilled water and condensing water flows, temperature rises, the condensate flow and temperature and steam pressure at the inlet to the chiller. Similar calculations could be made on other equipment such as a centrifugal chiller and process heat exchangers.

   This data will allow the operator to continuously monitor the performance of the prime mover and auxiliary equipment. For example it is not uncommon for a spark plug in a gas reciprocating engine to fail or the timing to change or the air-to-fuel ratio controller to get out of calibration. The system heat balance will immediately show an increased engine fuel consumption or increased exhaust gas temperature that should be corrected. An example of such a display is shown in Figures 4-1 through 4-5.

2. TOU electrical power produced by the prime movers should be data-logged for comparison with the electric bill. In order to do these calculations, the net energy consumption of the plant is calculated from TOU data. Consequently auxiliary power must be monitored on a TOU basis. Also carefully consider grouping of auxiliary loads related to power generation in a single motor control center if the cogeneration gas rate has a maximum monthly heat rate allowance. Placing the large loads associated with the thermal portion of the plant (chilled water pumps, condensing water pumps, etc.) on a separate load center will improve the net prime mover heat rate for cogeneration gas purchases.

3. Sufficient data must be logged for calculating the FERC efficiency on an actual basis. Utilities are considering or have started requiring daily electrical and thermal production reports to check the FERC efficiency of the plant. This suggests the general trend of utilities requiring more detail information and verification of actual plant performance.

Figure 4-1.

```
Saturday                    Le MERIDIEN RESORT                        Version:
01/21/89            MONITORING AND CONTROL SYSTEM                     01/16/89
14:22

                              Status

                    engine        absorber      centrifugal
-------------------------------------------------------------------
mode                optimum       optimum       manual
state               loaded        running       running
status              ok            ok            ok
process step        2             0             0
auto step           8             0             0
run time (h:m)      4173:36       3533:20       3175:02
-------------------------------------------------------------------
cause of shutdown:


     1          2          3          4          5          6
  CONTROL    DISPLAY      SET                             NEXT      Return
                                                                    EXIT
```

Figure 4-2.

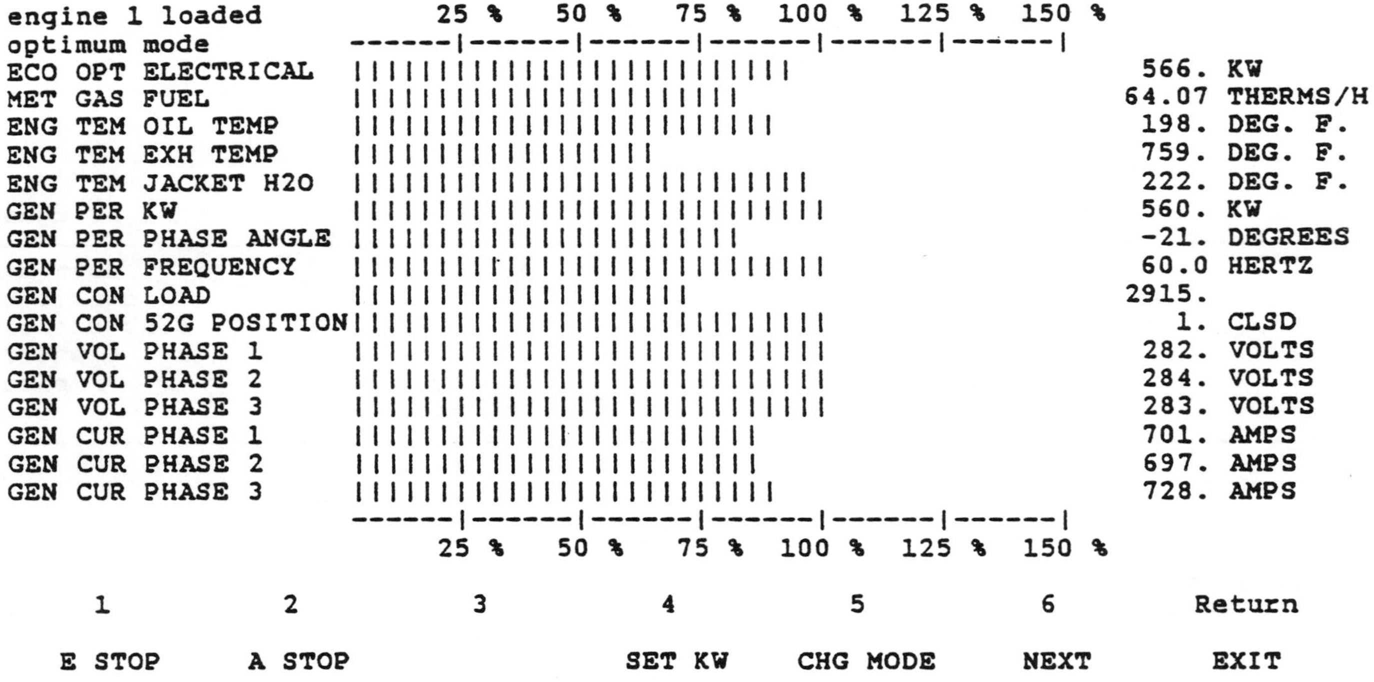

```
                          25 %  50 %  75 % 100 % 125 % 150 %
engine 1 loaded           |-----|-----|-----|-----|-----|
optimum mode
ECO OPT ELECTRICAL        |=================----|            566.   KW
MET GAS FUEL              |=============--|                  64.07  THERMS/H
ENG TEM OIL TEMP          |============|                     198.   DEG. F.
ENG TEM EXH TEMP          |===================|              759.   DEG. F.
ENG TEM JACKET H2O        |=================|                222.   DEG. F.
GEN PER KW                |================-|                 560.   KW
GEN PER PHASE ANGLE       |=================|                -21.   DEGREES
GEN PER FREQUENCY         |=================|                60.0   HERTZ
GEN CON LOAD              |================|                 2915.
GEN CON 52G POSITION      |======================|            1.    CLSD
GEN VOL PHASE 1           |=================|                282.   VOLTS
GEN VOL PHASE 2           |=================|                284.   VOLTS
GEN VOL PHASE 3           |=================|                283.   VOLTS
GEN CUR PHASE 1           |=============----|                701.   AMPS
GEN CUR PHASE 2           |=============|                    697.   AMPS
GEN CUR PHASE 3           |==============----|               728.   AMPS
                          |-----|-----|-----|-----|-----|
                          25 %  50 %  75 % 100 % 125 % 150 %

        1       2       3        4        5         6
      E STOP  A STOP          SET KW   CHG MODE   NEXT    Return   EXIT
```

Figure 4-3.

```
Point Group 13: economics          Saturday  01/21/89  14:11  Page 1
        25 %      50 %      75 %      100 %     125 %     150 %
        SAVINGS   PERFORMANCE ACTUAL
|ECONOMICS          SAVINGS    ACTUAL                      CA  20.70 $/HOUR
|ECONOMICS          PERFORMANCE PERIOD                     CA   2.
|ECONOMICS          PERFORMANCE PARASITES                  CA  30. KW
|ECONOMICS          PERFORMANCE NET GEN                    CA 334. KW
|ECONOMICS          PERFORMANCE HEATRATE                   CA 12653. BTU/KWH
|ECONOMICS          PERFORMANCE BOILER EFF.                CA  0.60
|ECONOMICS          PERFORMANCE KW/TON                     CA  0.90 KW/TON
|ECONOMICS          PERFORMANCE THM MUL'PLR                CA 1030. BTUS/FT3
        25 %      50 %      75 %      100 %     125 %     150 %
```

Figure 4-4.

```
Point Group 13: economics        Saturday  01/21/89  14:13  Page 3
       25 %      50 %      75 %      100 %      125 %      150 %
|--------|---------|---------|---------|---------|---------|
|ECONOMICS   GAS RATES      COGEN                         CA  0.3000 $/THERM
|=========================================
|ECONOMICS   AVOIDED COST   ELECTRICAL                    CA  16.05 $/HOUR
|***************
|ECONOMICS   AVOIDED COST   THERMAL                       CA  28.36 $/HOUR
|***************************************************
|ECONOMICS   AVOIDED COST   CHILLED H20                   CA   2.03 $/HOUR
|====
|ECONOMICS   AVOIDED COST   HTG HOT H20                   CA   1.67 $/HOUR
|===
|ECONOMICS   AVOIDED COST   SOFT HOT H20                  CA  26.69 $/HOUR
|=============================================================
|ECONOMICS   OPER. COST     FUEL                          CA  19.25 $/HOUR
|*****************
|ECONOMICS   OPER. COST     TOTAL                         CA  22.89 $/HOUR
|****************************************************
|--------|---------|---------|---------|---------|---------|
       25 %      50 %      75 %      100 %      125 %      150 %
```

Figure 4-5.

```
Point Group 13: economics    Saturday  01/21/89  14:15  Page 4
            25 %      50 %      75 %     100 %     125 %     150 %
|ECONOMICS OPER. COST MAINTENANCE||||||||                        CA 0.01000 $/KWH
|ECONOMICS LOADS      ELECTRICAL||                               CA   537. KW
|ECONOMICS LOADS      THERMAL|||||||                             CA  49.40 THERMS/H
|ECONOMICS PROJ'TD LOAD ELECTRICAL||||                           CA   531. KW/HR
|ECONOMICS PROJ'TD LOAD THERMAL|||||||||||||||||||||||||||||||   CA  11.87 THERMS/H
|ECONOMICS OPTIMUMS   ELECTRICAL                                 CA   228. KW
|ECONOMICS OPTIMUMS   THERMAL                                    CA  11.85 THERMS/H
|ECONOMICS CONTROL    MAX KW||||||||||||||||                     CA   600. KW
            25 %      50 %      75 %     100 %     125 %     150 %
```

4. To confirm utility billings consider requesting the utility to add a "pulse relay" to the TOU meter(s). This relay gives a contact closure based on the electrical power purchase/sale. The number of contact closures per unit time determines the instantaneous power. Integration of the signal over a period of time yields kW-hrs. This data allows detail checking of the utility bill and gives a precise basis upon which to base economic plant operating decisions.

5. Trend the performance of key components whose performance degrades with time. Examples include gas turbine compressor discharge pressure, inlet pressure to reciprocating engine, condenser pressures, etc. Recognition of these variables and determining when to take corrective action can significantly improve overall plant performance.

6. System reliability issues must be considered when designing a cogeneration system. Experience has shown that instrumentation failure seems to be the leading equipment based cause for a plant trip. Consider using the criteria that a single instrument failure should not shutdown the plant instead of the typical approach of merely using identical redundant instruments. The system should not be designed to trip (except for identified ESD requirements) unless there is first an alarm set such that there is sufficient time for the operator to take corrective action. Low reliability instruments should be considered for redundancy, either by monitoring both signals and using a software switch when one signal falls outside a specific range or by installing two identical instruments and wiring both to the termination cabinet. A single instrument failure can be corrected quickly by merely changing a single termination before the plant must shutdown.

7. We have observed many installations do not calibrate the instrumentation on an as-installed basis. RTD's and thermocouples have a significant voltage drop in the plant wiring. An error of 10 F or more is not unusual if the instrument is not calibrated in place. This temperature error may be significant when monitoring steam temperatures, under or overstating thermal billing data or at least misleading to the operator. Incorrectly calibrated instruments will also preclude calculating an accurate plant heat balance.

8. Consider designing the system such that the key operating parameter data are frozen at the instant the plant trips and stored in a separate time stamped data buffer. Such a data log is invaluable when trying to sort out primary and consequential trips. A first-out annunciator will not give this key information as it only tells us what actually failed. Many times consequential failures are logged first over the primary failure mode. An example of such a log is shown in Figures 4-6 through 4-8.

9. Key performance data such as utility billing data, long term equipment performance trends, customer primary and secondary billing data, etc. should be permanently logged either locally or off-site.

10. Many plants are considering a used backup generator set solely to avoid demand charges. An emergency generator set is usually exempt or comes under a category of engines more easily licensed, is relatively inexpensive and can pay for itself in less than a year by avoided demand charges. A 500 kW demand avoided for 12 months at $10.00/kW/month is $60,000.

## FUTURE TRENDS

Many utilities, especially in California, depend upon cogenerated power as an integral part of their current and future resource plans. Problems still exist in the view of the utility controlling several dispersed power generators, controlling off-peak system stability with a large percentage of cogenerated power, and projecting cogeneration system reliability. Cogenerators view the interconnection issues differently relying on the utility as a standby electrical energy source and as a sink for any surplus electrical energy.

A conceptual approach to bringing these diverse problems closer to solution has been proposed. A cogeneration monitoring and control network would interconnect each cogenerator directly with the serving utility. This network would be used to continuously transfer selected plant operating data to the utility for instantaneous and future system resource and reliability planning.

In addition such a network would integrate the dispersed power generation of cogenerators into the utility system by providing perfor-

**Figure 4-6**

ERROR AND CRITICAL PARAMETER LOSS

Following is an example of a chronological error log. The error log should contain all alarm and change-of-state information for the facility.

| 01/21/89 | 06:25 | Chiller | Running | | |
|---|---|---|---|---|---|
| 01/21/89 | 06:20 | SPACE HTG. | TEMPERATURE | HOUSE | 121.7 DEG. |
| F. warning | | 19/ 3 TE-7-3 | | | |
| 01/21/89 | 06:20 | Chiller Starting | | | |
| 01/20/89 | 16:43 | Chiller Stop | | | |
| 01/20/89 | 13:27 | ENGINE | LEVEL | OIL TANK | |
| restored | | 16/ 2 LSL -11-1 | | | |
| 01/20/89 | 03:41 | Chiller Starting | | | |
| 01/19/89 | 20:35 | Chiller Stop | | | |
| 01/19/89 | 05:53 | Chiller Starting | | | |

Another type of log which is extremely useful in trouble shooting problems after a shutdown or trip has occurred is a history log of critical performance parameters. This log should log these parameters approximately each 15 seconds. When a shutdown or other fault occurs, the log is frozen. The log may then be studied by the operator to help in determining the cause of the problem.

Following is an example of what a log of this type may look like:

```
History Log Report          PALOMAR COLLEGE          01/21/89          13:53
Page 1

01/21/89 13:53:47
SYSTEM            VOLTAGE        PHASE A TO B                         480.
VOLTS ENGINE
77. PSIG ENGINE    PRESSURE       OIL
34.4 PSIG ENGINE   PRESSURE       GAS
9.1 PSIG ENGINE    PRESSURE       RT. IN. MAN. PE-11-1
10.8 PSIG ENGINE   PERFORMANCE    LT. IN. MAN. PE-11-2
15.58 THRM/HR ENGINE TEMPERATURE  THERM OUTPUT
184. DEG. F. ENGINE  TEMPERATURE  OIL             TE-10-1
-1400. DEG. F. GENERATOR VOLTAGE  EXHAUST         TE-2-1
476. VOLTS GENERATOR  VOLTAGE     PHASE A
476. VOLTS GENERATOR  VOLTAGE     PHASE B
480. VOLTS GENERATOR  CURRENT     PHASE C
475. AMPS GENERATOR   CURRENT     PHASE A
470. AMPS GENERATOR   CURRENT     PHASE B
470. AMPS GENERATOR   ELECTRICAL  PHASE C
-14. DEGREES GENERATOR ELECTRICAL PHASE ANGLE
59.9 HERTZ GENERATOR  PERFORMANCE FREQUENCY
                                  POWER
```

**Figure 4-7.**

01/21/89 13:53:30

| SYSTEM | VOLTAGE | PHASE A TO B | 480. |
|---|---|---|---|
| VOLTS ENGINE | | | |
| 78. PSIG ENGINE | PRESSURE | OIL | |
| 34.4 PSIG ENGINE | PRESSURE | GAS | |
| 9.3 PSIG ENGINE | PRESSURE | RT. IN. MAN. PE-11-1 | |
| 11.1 PSIG ENGINE | PRESSURE | LT. IN. MAN. PE-11-2 | |
| 14.76 THRM/HR ENGINE | PERFORMANCE | THERM OUTPUT | |
| 184. DEG. F. ENGINE | TEMPERATURE | OIL | TE-10-1 |
| -1400. DEG. F. GENERATOR | TEMPERATURE | EXHAUST | TE-2-1 |
| 477. VOLTS GENERATOR | VOLTAGE | PHASE A | |
| 476. VOLTS GENERATOR | VOLTAGE | PHASE B | |
| 480. VOLTS GENERATOR | VOLTAGE | PHASE C | |
| 472. AMPS GENERATOR | CURRENT | PHASE A | |
| 468. AMPS GENERATOR | CURRENT | PHASE B | |
| 467. AMPS GENERATOR | CURRENT | PHASE C | |
| -14. DEGREES GENERATOR | ELECTRICAL | PHASE ANGLE | |
| 59.8 HERTZ GENERATOR | ELECTRICAL | FREQUENCY | |
| 376. kW GENERATOR | PERFORMANCE | POWER | |
| -94. KVAR | PERFORMANCE | KVAR | |

01/21/89 13:53:13

| SYSTEM | VOLTAGE | PHASE A TO B | | 480. |
|---|---|---|---|---|
| VOLTS ENGINE | | | | |
| 77. PSIG ENGINE | PRESSURE | OIL | | |
| 34.3 PSIG ENGINE | PRESSURE | GAS | | |
| 9.6 PSIG ENGINE | PRESSURE | RT. IN. MAN. PE-11-1 | | |
| 11.3 PSIG ENGINE | PRESSURE | LT. IN. MAN. PE-11-2 | | |
| 14.94 THRM/HR ENGINE | PERFORMANCE | THERM OUTPUT | | |
| 183. DEG. F. ENGINE | TEMPERATURE | OIL | TE-10-1 | |
| -1400. DEG. F. GENERATOR | TEMPERATURE | EXHAUST | TE-2-1 | |
| 476. VOLTS GENERATOR | VOLTAGE | PHASE A | | |
| 477. VOLTS GENERATOR | VOLTAGE | PHASE B | | |
| 480. VOLTS GENERATOR | VOLTAGE | PHASE C | | |
| 489. AMPS GENERATOR | CURRENT | PHASE A | | |
| 482. AMPS GENERATOR | CURRENT | PHASE A | | |
| 478. AMPS GENERATOR | CURRENT | PHASE C | | |
| -15. DEGREES GENERATOR | ELECTRICAL | PHASE ANGLE | | |
| 59.8 HERTZ GENERATOR | ELECTRICAL | FREQUENCY | | |
| 387. KW GENERATOR | PERFORMANCE | POWER | | |
| -104. KVAR | PERFORMANCE | KVAR | | |

**Figure 4-8.**

```
01/21/89 13:52:55
SYSTEM              VOLTAGE
VOLTS ENGINE        PRESSURE      PHASE A TO B
80. PSIG ENGINE     PRESSURE      OIL
33.5 PSIG ENGINE    PRESSURE      GAS
9.8 PSIG ENGINE     PRESSURE      RT. IN. MAN. PE-11-1
11.4 PSIG ENGINE    PERFORMANCE   LT. IN. MAN. PE-11-2
14.76 THRM/HR ENGINE              THERM OUTPUT
183. DEG. F. ENGINE               TEMPERATURE    OIL        TE-10-1
                                  TEMPERATURE EXHAUST       TE-2-1

-1400. DEG. F. GENERATOR
476. VOLTS GENERATOR              VOLTAGE        PHASE A
477. VOLTS GENERATOR              VOLTAGE        PHASE B
481. VOLTS GENERATOR              VOLTAGE        PHASE C        481.
496. AMPS GENERATOR               CURRENT        PHASE A
488. AMPS GENERATOR               CURRENT        PHASE B
485. AMPS GENERATOR               CURRENT        PHASE C
-14. DEGREES GENERATOR            ELECTRICAL     PHASE ANGLE
59.9 HERTZ GENERATOR              ELECTRICAL     FREQUENCY
391. KW GENERATORPERFORMANCE      PERFORMANCE    POWER
-97. KVAR                         KVAR
```

mance and status information as the utility requires. The network would satisfy the reporting needs of the utility while providing uniformity and continuity in information flowing from each cogenerator.

The utility also benefits by the ability to dispatch individual and groups of cogeneration plants if it is to the economic benefit of the cogenerator and the utility. Finally, the cogenerator and the utility would certainly develop a working relationship on the operational level that benefits both organizations in the long run.

The cogenerator benefits from the network by having an automatic reporting and processing of system performance information to the owner, operator and the serving Utility. The network is also an interface to the serving utility which would handle all reporting requirements. The network would track and report thermal efficiencies to assure compliance with FERC regulations and utility contract requirements.

Continual and instantaneous electrical billing and sale data would also be available from the network for the cogeneration. Perhaps most importantly it provides the cogenerator the potential for increasing electrical sale revenues. If is in the best interests of the utility to reduce cogenerated power during the off-peak period, then the utility could dispatch certain plants accordingly, assuming the appropriate economic incentives are made available to the cogenerator.

Conversely, if the utility requires additional electrical power during the summer peak, the network could be used to purchase additional economically attractive surplus power from cogenerators.

There are surely many pitfalls that must be overcome before a network as described becomes a reality. Utilities and cogenerators may view such a proposal as intrusive and unnecessarily burdensome. However the potential benefits to the cogenerator and the utility require each to examine such an approach to improving the cogeneration and utility system economics.

*Chapter 5*

# Risk Management Issues
# For Cogenerators and IPPs

# Risk Management Issues For Cogenerators and IPPs

*Michael P. Golden*
*Marsh & McLennan, Inc.*

**M**anaging the risk for an Independent Power Project or Cogeneration facility is a very challenging task. Market instability, "spotty loss history" and geographic diversity conspire to make this a somewhat challenging placement. This chapter will explore some of the key risk management issues related to the operation of these assets in an environment that is undergoing restructuring.

Insurance (by itself) is not the answer to concerns about negotiating power sales contracts in today's restructured world. The unique aspects of power purchase contracts do make it difficult for underwriters, even those that are comfortable with the risks associated with your regulated brethren. Therefore, you and your insurance and risk management advisors must approach the problem and the markets in a more deliberate and careful manner.

While most of this discussion will be focused on private sector domestic projects, most if not all of the issues covered (and then some) are applicable to foreign operations as well. The perspective taken here is that of the party responsible for developing the projects and overseeing the ongoing operation.

The objectives in this chapter are to:

1.  Develop a fundamental understanding of the risk management process for independent power and cogeneration operations;

2.   Clarify the differences between "risk management" and "insurance,"

3.   Identify the risks associated with a non regulated power project;

4.   Outline a typical "insurance" policy for a power project,

5.   Discuss several key issues associated with risk management, the insurance market place and the non regulated power industry.

# FUNDAMENTALS OF RISK MANAGEMENT

It is important to establish the paradigm that my colleagues and I use when looking at risk. If one accepts the premise that uncertainty is not good, then it is in the best interests of the enterprise to minimize the probability of occurrence of those events or situations that contribute to uncertainty. The "old school" looked at risk from the perspective of available insurance products. Fire insurance is available, therefore, the fire risk is "taken care" of by insurance. If a certain risk was not insurable, then it was essentially ignored.

Today's risk and insurance manager looks to the "circle of risk management" as reference for dealing with the risks and exposures of the business. The "circle" analogy is appropriate because there is neither an end nor a beginning to the management of risk. As long as there is internal and or external changes, there will be the need to continually "go around" the risk management circle.

Let's first establish the components of the circle. Throughout the remainder of this discussion, I will highlight examples of how these concepts are or should be applied.

Obviously, the first step is the *identification* of all risks. This step should not just be limited to "insurable" risks. If the organization is interested in quantifying its total risk quotient, this process must go beyond the traditional areas of property, boiler and liability. Surveys, research, benchmarking and "creative thinking" are the ways we use to identify risk. The key to postulating maximum foreseeable and probable maximum losses (sorry for the industry jargon) is to not allow the engineers and operators "low-ball" the potential for loss. Protective systems and base designs work, but there always seems to be that exception that happens to the "other guy."

After inventorying and quantifying the risks, management needs to determine which exposures should be totally avoided. If you and your financial partners are not comfortable with the California earthquake exposure—avoid this risk and locate your plant in a more stable part of the country. While often the least expensive of the steps, there could be significant missed opportunity costs associated with the avoidance strategy.

We are all risk takers. Therefore, it is not necessarily a natural reaction to avoid risk. We are more likely to seek ways to maximize our return while still taking some risk. To make certain that this is not always a zero sum game, the prudent risk management professional places great emphasis on the minimization of risk. Unfortunately, all too often, safety and loss control are considered expense items. Often subject to the vagaries of expense control and relegated to non operational organizations, these professionals do not always have the influence and impact that they should.

Today's loss control and safety specialists do more than just count sprinkler heads or monitor the wearing of steel toed safety shoes. They are dedicated to minimizing the occurrences of loss. They consider both pre- as well as post-event loss scenarios and try to develop systems, procedures and processes that either reduce the frequency or occurrence of all foreseeable loss situations. Having sufficient under turbine fire protection, adequate spare parts and an effective crisis management program are examples of what should be key components of an integrated risk management program for every IPP-cogen facility.

There is a point at which the entity must consciously assume uncertainty. Perhaps the easiest and least expensive way to deal with risk is to transfer it contractually. Readers are well aware of the power of the contract and the assumption of risk. Your financial backers want to assume no risk and seek to transfer it all to you. As long as there is a balance between the risks that are assumed via a contract, transferred via insurance and clearly retained, the entity can forecast its cost of risk. When one side permits risk "leakage" or an imbalance to occur, an unmanaged exposure exists that could jeopardize the profitability, or even viability of the organization.

Risk financing is the remaining "spot" on our wheel. Risk financing includes two components—risk retention and risk transfer (or as most of us know it—insurance).

The risk retention option, or self insurance or self funding, seeks to duplicate what the underwriter does while avoiding the frictional costs associated with an insurance policy. Self insurance is the follow-on to the adage: "do not insure for those events you know you are going to have." While self insurance does not have the underwriter's advantage of true risk spreading, the financially strong firm can take a long term view of frequency exposures and make allowances for them as a routine cost of doing business. Transferring risks such as automobile physical damage (fender benders) or small component breakdowns is simply not cost effective. Assuming the risk of loss for a low frequency, high severity event such as a flood is not recommended.

Risk retention takes on several forms. It can be as simple as establishing a reserve account for worker's compensation claims to as complex as reinsuring your own captive insurance company with a finite risk contract. No matter what the form, the retained portion of your risk portfolio should not include any surprises and to the extent possible have built in protection for "shock" and batch losses.

Through careful analysis, peer studies, and some guesswork, insureds, brokers and underwriters seek to apply the "law of large numbers" and determine an efficient "premium" for transferring the risk. Since insurers are (or are supposed to be) profit generating organizations, the market is fluid and not always predictable.

Insurers seek to spread their exposures over many insureds. They further attempt to protect themselves by transferring a portion of their risk to reinsurers. Insurers' interest in classes of risk such as IPPs and cogen projects sometimes vary with the short term profitability of that class. A "bad" year combined with a less than impressive investment portfolio can often be enough to dramatically push a market out of a particular class.

Sometimes, an insurer will think they are smarter than the rest and "buy" market share with below average prices. This results in the "Roman Candle" effect—a quick start, an explosion and then, nothing. They are out of the market just as quick as they came in because they failed to understand the risk, failed to appreciate the exposure and failed to accept the very long term implications of your business.

For most, a combined risk retention-risk transfer strategy is the most appropriate. You and your advisors and brokers should seek to structure a program that is stable, cost effective and responsive to the business issues.

However, note that the ideal program is the culmination of the five step approach. The most successful risk management programs contain each one of the elements of our circle and subjects the program to a constant journey around the circle.

# THE RISKS

The risks associated with a private sector non regulated power project can be categorized under numerous broad headings. For purposes of this chapter I will distinguish between non-project and project risks.

### Non Project Risks

Non-project risks are those not directly related to the project. They include:

1.  Adverse movements in exchange rates;
2.  Interest rate fluctuations;
3.  Inflation;
4.  Political instability;
5.  Changes in government regulations and similar risks.

As a general rule, these risks are not insurable by traditional methods.

### Project Risks

Project risks are those directly related to the project. They can be further distinguished between "commercial risks" and "fortuitous risks." Commercial risks are, in essence, trading risks and can include:

1.  Loss of tender or contract to a competitor
2.  Failure to obtain necessary permits
3.  Incorrect pricing of the cost of power
4.  Lack of demand for electricity

Commercial risks are generally not insurable although in special circumstances certain commercial risks can be and have been insured. For instance, insurance has been arranged for the loss of developmental expenditures following the failure to obtain permits or the failure of certain international treaties to be ratified. However, it must be pointed out, that these are the exceptions rather than the rule.

Fortuitous risks are those happening by chance or accident and include:

1. Physical loss or damage and the resultant loss of income or profits;
2. Injury to employees or third parties;
3. Defects in design, materials or workmanship;
4. Events of force majeure;
5. Delay in completing construction;
6. Interruption to commercial operations;
7. Consequences of professional negligence,
8. Dishonesty of employees;
9. Machinery breakdown,
10. Management liability.

These risks are generally insurable subject to underwriting terms and conditions that ensure spread of risk and the elimination of moral hazard concerns.

# THE TYPICAL INSURANCE PROGRAM

Just as there is no one standard IPP-cogen "deal," there are no standard insurance and risk management programs for this industry.

A number of parties have an interest in the risk management and insurance programs associated with a power project. In addition to the lenders and developers, the government, contractors, suppliers and the public have legitimate insurable interests in the project. During the life of the project these interest classes will change in importance and some-times have conflicting views. In order to adequately satisfy the different interests there needs to be focused coordination of the risk and insurance strategies and central purchasing of coverages and supporting services.

However, it should be readily apparent that the financial backers of the operation are the ones that often dictate the terms and rigidly enforce their compliance. Unfortunately, however, there are often significant gaps between the financial institution's expectations of the insurance market's appetite for IPP-cogen project risk. Failure to identify this "expectation deficit" can lead to serious problems later on.

This certainly means that the professionals negotiating the financing and power purchase contracts and the professionals negotiating the insurance and risk management contracts need to work closely together so that the balance of retained, assumed and transferred risk we spoke about earlier is maintained.

Power project insurance programs are generally arranged in two phases. The first covers the construction phase, including any testing and commissioning periods. The second covers the commercial operating phase and is generally arranged on an annually renewable basis.

### The Construction Phase

Either the contractor or the owner/developer assumes responsibilities for arranging and purchasing insurance during the contract phase. Among the insurance contracts that are arranged during this phase are:

1.  *Construction all risks insurance:* covers physical loss of or damage to the construction works or other property while at the site or during inland transit.

2.  *Delay in start-up insurance:* covers the financial consequences of delay in commencement of commercial operations caused by physical loss or damage to the works.

3.  *Marine Cargo Insurance:* covers physical loss of or damage to equipment and other supplies transported by air or sea.

4.  *Marine delay in start up insurance:* covers the financial consequences of delay in commencement of commercial operations caused by a physical loss, damage or disappearance during marine transportation.

5.  *Third-party liability insurance:* covers the legal liability of all parties arising out of bodily injury to or property damage of third parties

Each party involved in the construction project will also typically arrange for policies or fund for the following exposures:

1.  Workers' compensation or employers liability, as appropriate for the jurisdiction;

2.   Design engineers, architects or other professional consultants errors & omissions as prescribed by the contract;

3.   Contractors tools and equipment;

4.   Automobiles;

5.   Employee dishonesty, fiduciary and management liability (D&O) exposures.

**The Operating Phase**

For the operating phase, responsibility for developing a risk management strategy and arranging and placing the insurance clearly lies with the owner-operator of the facility. The coverages here are essentially the same as in the construction phase except for the need to address the machinery breakdown and business interruption exposures. Unfortunately, these are the two areas that tend to be the most difficult for us to deal with.

As mentioned earlier, the lenders have a keen interest in the insurance program from "day one." They want an assurance that the borrower can service the debt—or in other words, insurance against the financial consequences of delay in completion of construction or interruption of commercial operations.

The extent to which insurance is, or can be, purchased is influenced by the willingness of the worldwide insurance and reinsurance markets to provide this cover. While this is a somewhat gratuitous statement in general terms, with respect to the IPP and cogen time element exposures—delay in start-up and business interruption, this willingness or unwillingness is very much conditioned by the ability of the insured and broker to adequately explain the exposure to the underwriters.

# ISSUES

In assessing Marsh & McLennan's experience in dealing with IPP-cogen risk management programs, a number of "issues" become apparent. The following discussion is not meant to be an exhaustive analysis of all the insurance and risk management trends, but rather a brief overview of some of the more critical challenges we all must deal with.

**Business Interruption Exposure**

When the insurance industry speaks of "business interruption," in its basic form, it means insurance to do for the insured during a period of business interruption what the business would have done had no loss occurred. Loss of business earnings, the prime source of money for continuing operating expenses as well as profit, is the subject of this coverage. It is important to note that under the traditional form, recovery applies only for the time required to repair or replace damaged property with "reasonable speed," so that normal operations can be resumed.

In the IPP-cogen world, the basics of business interruption are slightly more complex. Disagreements and confusion over the business income or business interruption exposure is perhaps the leading cause of insurer disinterest in the IPP cogen exposure.

Because your purchase agreements provide for availability clauses and incentives, the actual period of loss can extend well beyond the repair period. Underwriters and claims adjusters, if not completely familiar with the terms of the contract, will resist accepting the incentive loss (or penalty). Another area of confusion involves the loss of capacity payments caused by a failure during a power purchaser availability test. If the policy has the usual daily deductible for business interruption, the claim could be denied because the plant had not met the specified waiting period.

*Solution*

The key to placing the proper business interruption program is communication. It is essential that your risk manager and risk advisor/ broker have access to those negotiating the power purchase contracts. It is suggested that you not only discuss the terms of the power purchase agreement with the risk management team, you should also insist that various loss scenarios be "gamed out" to verify that everyone understands the various triggers, clauses and incentives contained in the contract. With knowledge, your broker can then seek to structure a program that provides you the most efficient protection.

Armed with a thorough understanding and/or the ability to bring the IPP-cogen contract professionals to meet the underwriter, we can attempt to reduce the deductible waiting periods, substitute a financial deductible for a daily deductible, amend the indemnity measure to reflect the true loss and or extend the indemnity term beyond the outage period.

## Marine Delay in Start Up

If a major piece of equipment is being shipped by air or sea, the lenders may insist that marine delay in start up coverage be purchased. Often, the component manufacturer will assume responsibility for purchasing the cargo coverage on the shipment. This therefore leaves the IPP-cogen developer in the position of having to purchase a "mono-line" coverage at very, very steep rates.

*Solution*

The most obvious solution is to try to package the cargo and marine delay in transit coverage. Assuming that the proper credits can be obtained from the manufacturer, this is an effective solution.

Perhaps an even simpler solution is to assess the exposure and explain it to the bankers. For example, assume the bank requires the purchase of this insurance for the shipment of a generator. If the generator is not on the critical path for commercial operations and you can demonstrate that a spare can be located and shipped in time to meet the projected start date, the need for insurance or at least the initial required limits, should be minimized. This is simply an example of employing the risk management techniques we discussed above.

## Underwriter's Interest

Insurers of "technical risks" such as IPP-cogen facilities expect and need detailed engineering data to properly underwrite the account. It is our experience that engineering documents detailing maintenance histories and procedures, fire protection system capabilities and operating logs are not always available. Furthermore, some underwriters have the impression that loss control and safety is not a priority for the IPP-cogen industry. Underwriters are very aware of the fact that since 1990, there have been at least four fires at non regulated units where the physical damage alone has ranged from $4,000,000 to $23,000,000.

While I would not necessarily agree with some underwriter's low opinion of IPP-cogen loss control, our inability to consistently present, as pan of our underwriting submission, a satisfactory explanation of the property protection and safety control programs, does not allow us to remove that uncertainty from the underwriter's premium calculations. The concern for loss control goes beyond premiums to insurability. Whether it is an older component with a history or a new unit that may be of a suspect design, underwriters expect and need to see evidence of your commitment to preserving equipment and preventing loss. In other words—more information is better.

## The Marketplace

As briefly discussed earlier, the IPP-cogen sector has seen a variety of insurers enter, leave and re-enter the market. Of those that have remained in the game and evidence a longer term approach to this sector are:

1.  American International Group (AIG)
2.  Hartford Steam Boiler (HSB)
3.  Reliance
4.  CIGNA
5.  AEGIS
6.  Chubb

While none of these markets is necessarily a "one stop" shop for all your needs, they have over the past decade demonstrated the ability to profitably underwrite their products for your risks. Again, the key to accessing these markets is to seek a partnership relationship—one that encourages open communication and a willingness to seek long term commitments even at the expense of short term savings. The strategy of chasing price reductions does nothing but "burn out" the market.

Your financing contracts and power purchase agreements typically span a number of years. Even though the above mentioned markets have been long-term players, their willingness to provide particular coverages, grant certain deductibles or provide specified limits has changed over the years. Therefore, if your contracts contain very explicit insurance requirements with respect to specific coverages, limits or retentions, it is quite possible that over the course of the contract we will face a market cycle that "blocks" the availability of the required coverages.

As an additional complication, the financial strength of an insurer is a concern to everyone. Often your lenders will require that you purchase insurance from only "A" rated companies. While this is an admirable goal, it may not be realistic, given the long term of the contract.

*Solution*

Attempt to secure as flexible insurance terms as possible in your contracts. However, this does not mean, ignore insurance until the last moment. Consider adding a provision such as "as available on reasonable terms and conditions" to allow you to adjust to market cycles. Some insurers are willing to consider multiple year programs. These can be beneficial as long as the cancellation clauses are understood on both sides.

**Political Risk**

There have been numerous papers, speeches and conferences dedicated to the political risk exposure for IPP and cogen projects. It is not my intention to deal with this topic in any detail except to say that the exposure is real and capacity is limited.

*Solution*

Your risk management professional must play an integral role in assessing the country risk component of the project. Political risk insurance is by no means a "standard" product. It is important to engage a specialist in this field and customize a program that responds to your own individual risk profile.

# CONCLUSION

The convergence of utility industry deregulation, privatization, competition, global growth and the advance of power generation technology suggest that the IPP-cogen "players" have a challenging but very bright future. This future, as bright as it might be, is not isolated from risk and uncertainty. The professional risk management community is prepared to work with you—the developers, contractors, suppliers and financiers—to contain this risk and provide the "bottom-line" protection you require.

*Chapter 6*

# Managing Risks During the Construction of a Cogeneration Facility

*Chapter 6*

# Managing Risks During The Construction of a Cogeneration Facility

*Michael C. Loulakis, Esquire*
*Wickwire, Gavin, P.C.*

T he construction of a power generation facility is a substantial undertaking—involving considerable risks to all parties involved. While contractors are accustomed to dealing with risks, construction owners are typically more naive about not only the risks they are assuming in the construction of a project, but also about the role they play on the project itself.

Owners and developers of power facilities must understand at the outset that their role during the construction of a project is as integral to the success of the project as that of the designer and contractor. In addition, owners should also understand that there are virtually no risks on a construction project that cannot be shifted among the contracting parties as part of the business deal. Consequently, an owner may contractually be assuming the risks of (1) unusually severe weather, (2) unexpected subsurface conditions, (3) strikes at the turbine supplier's plant or (4) changes in law—as well as the increases in price and delays to project completion associated with such risks.

In light of this, a prudent owner will evaluate more than just whether there is sufficient financing to complete the construction of a contemplated project. Prudent owners will conduct a risk management review of the project structure and the contracting terms, with the primary focus being (1) the identification and analysis of the most significant risks faced, (2) a determination of how such risks can be either

mitigated or eliminated, and (3) the assessment of the financial exposure
to the owner should the potential risk become a reality. This chapter will
present the framework that owners and developers of power generation
projects can use in undertaking such a risk management review.[1]

# UNDERSTANDING YOUR OWN STRENGTHS AND WEAKNESSES

One of the most frequently overlooked risks that you must face, if
you are a construction owner, is whether you have the requisite experience to build the project you are contemplating. Some of the specific
questions that you might consider are the following.[2]

1.  *Is the project a natural outgrowth of your previous experience?* From a
    risk management perspective, you should not assume that your
    experience in one area of the power generation industry gives you
    the experience you need to be successful in other areas. For ex-
    ample, hydroelectric facilities, cogeneration plants and waste-to-
    energy plants all involve different technologies, different sets of
    experienced contractors and different regulatory issues that must
    be understood before starting the contracting process.

2.  *Is your personnel experienced?* You should have a strong sense of the
    technical capabilities of your staff and whether your staff is capable
    of constructing the project you are considering.

3.  *Are you familiar with the location of the project?* As experienced con-
    struction owners know well, the success of a project can be affected
    by local conditions. Local labor conditions, interpretation of build-
    ing codes, weather conditions and citizen activism are all issues
    that must be evaluated.

If you have answered any of these questions in the negative, this
does not mean that you would be imprudent in going forward with the
project. It does mean, however, that you will have to take great care in
assembling a project team that will be capable of overcoming your weak-
nesses and lack of experience.

# ANALYZING THE PROJECT'S ORGANIZATIONAL STRUCTURE

One of the first steps an owner must take in going forward with a cogeneration project is to determine the organizational structure that will govern the relationships among the parties on the project—owner, architect-engineer, contractor and subcontractors. A well-planned organizational structure will greatly enhance the success of the project and help avoid many of the risks and conflicts involved in construction contracting. Conversely, the lack of strategic thinking will assuredly make the project more difficult, increase the likelihood of conflict and reduce, or even preclude, the project's chances for success.

There are a variety of contract delivery methods available to an owner building a construction project. Each has their peculiar risks and rewards to the owner—and must be chosen with a great degree of care to ensure that the overall goals of the owner are being achieved. Set forth below are the most common contracting methods used in the construction of power generation facilities.

## Traditional General Contracting

Under a traditional method of contracting, the owner hires an architect or an engineer to specifically design the entire project.[3] When the design is completed, the owner hires a general contractor to construct the project according to the plans and specifications prepared by the designer. The contractor may choose to hire trade subcontractors and suppliers to perform portions of the work. During construction, the designer and owner monitor the work to insure that the contractor precisely follows the plans and specifications.

Because of the technical complexity of most power generation facilities, it is often impractical to use the traditional method of contracting. Owners want to have a facility that is guaranteed to achieve certain objectives related to output and emissions. Design professionals do not provide such guarantees in formulating designs, partly because this is outside the scope of their insurance coverage. On the other hand, the traditional general contractor does not provide such a guarantee—simply warranting that it will construct the project in accordance with the design tendered by the design professional.

Another concern is that because of the fast track nature of the construction of an energy project, it is often difficult to have a design complete before construction starts. As a result, a contractor might be asked to bid off a set of plans and specifications that may be less than 100% complete, thereby resulting later in disputes over whether items specified in the final design should have been assumed by the contractor in the original bid. This can result in major claims at the end of the project.

## Design-Build

As a means of resolving the issue of guarantee, most power generation projects are built under the design-build form of contracting.[4] Design-build—also known as turnkey and EPC (Engineer, Procure and Construct) contracts—calls for one entity, known as the design-builder, to undertake the responsibility for both the design and the construction of the project.

Power generation facilities are particularly appropriate for a design-build contract delivery. First, design-build enables the owner to hold one party accountable both for the design and construction of the entire project. A design-build approach brings the construction project much closer to being a product than under the traditional methods of contracting. This single-point contact not only relieves the owner of the need to coordinate the designer and the contractor—one primary cause of construction disputes and overruns—but also enables the owner to specifically contract for performance guarantees relating to project.

It should be noted that there are several risks that an owner assumes when using the design-build contracting approach. For example, although the owner derives the benefit of having one party responsible for the complete development and construction of a project, the owner must rely solely upon that party for any recovery of compensation if something goes wrong. To counter this risk, many owners ask for financial guarantees or bonds from third-parties so as to insure that there is substance behind the construction organization. Others seek equity participation by the design-build entity as a means of ensuring proper project performance.

Another risk is that the design-build method eliminates the checks and balances that are present when design and construction are separate. Under the traditional approach, design professionals closely examine a general contractor's performance to determine whether it meets specifications and justifies payment. No such checks and balances exist when the design and construction are being done through one entity. Prudence suggests that an owner have its own in-house staff, or an outside engineering firm, review the work of the design-builder and insure that the product that is being furnished to the owner meet the owner's objectives.

Finally, it should be remembered that a *true* design-builder is one who takes full responsibility for design. Some developers attempt to mitigate costs by performing large portions of the design in-house, and then executing a contract that purports to be design-build. This can lead to arguments over whether the design-builder was merely completing the design, based upon assumptions of the developer, or was fully certifying the adequacy of the entire design. In this situation, should the design-builder be responsible for extra costs arising out of that portion of the work designed by the developer? The design-build contract should resolve this issue clearly.

### Identify the Contract Delivery Method

A common oversight by an owner is to inadequately describe the contract delivery method that will be used on the project. For example, if the contractor will be acting as a design-builder, care must be taken to fully describe the design functions the contractor will be undertaking. Owners who fail to specify this in the contract may face an argument during construction that the contractor's design responsibilities were more limited—perhaps simply to reviewing the owner's performance requirements—than the owner originally intended. The owner should also specify what obligations the contractor will have for start-up, testing, operation and maintenance on the project.

# PROTECTION THROUGH THE CONTRACT

In assessing who should bear the responsibility for construction risks, it is critical to remember that virtually any risk can be assumed for the right price. Consequently, one of the most important functions of a construction contract is to properly allocate the rights, responsibilities and risks assumed by the parties to the contract.

Owners should also understand that the goal of sound risk management is not to shift the risk of every unknown to the contractor, since this will likely result in inflated and unreasonable base contract bids. The goal of the prudent construction owner is to determine what risks it can live with, structure such risks into the business deal and reflect such risk allocation in the contract.

## Changes

One of the major risks that an owner must recognize is that construction projects such as power generation facilities are rarely completed in precisely the same fashion contemplated by the parties at the time of the agreement. This is because the circumstances and conditions which delineate the scope and parameters of a given project vary over time. In most cases the parties have little or no control over these changing conditions.

Changes may be required for many reasons, including:

1.  *Third-party requirements.* This may come in the form of governmental regulatory agencies or changes in the law.

2.  *Owner's changing requirements.* For example, if the owner of a cogeneration project is also the host facility and decides to expand its manufacturing plant, it may need to change the project to obtain additional thermal output to support the plant expansion.

3.  *Changes in technology.* This may occur if use of a particular process becomes less economical than recently developed alternatives. Also, situations can arise where the technology specified by a contract is not compatible with actual conditions encountered.

Astute owners should insist upon contractual provisions which give them the flexibility to secure acceptable changes in the scope of work for a reasonable price. The Changes clause of the contract typically allows an owner to direct unilaterally changes in (1) the specifications, (2) the manner or method of performance, (3) the time of performance, or (4) the equipment, materials, facilities or services provided by the owner.[5]

It is important for an owner to recognize that one of the most important points about the Changes clause is the right to direct a change without the contractor's consent to the change. This allows the change to be made without giving the contractor the right to insist upon unreasonable time or money concessions as a condition to performing the work. Nevertheless, the obvious risk to the owner is that by proceeding with the change in the absence of an agreement on price and time, the owner will face the risk of actual contractor costs being higher than the owner might have anticipated. Thus, even though the contract should give the owner the right to make unilateral changes, it is prudent to (1) discuss the change in depth with the contractor, (2) give the contractor adequate time to price the change and integrate the change into the work, and (3) to reach a lump sum price before the change is performed.

The Changes clause should contain a formula for determination of an appropriate equitable adjustment in the event the contractor is directed to proceed prior to an agreement on price or schedule. Generally, the contractor is entitled to be compensated for its additional costs together with a reasonable allowance for overhead and profit. Owners should give some consideration to limiting overhead and profit to a percentage of direct costs. To ensure no misunderstanding, the elements comprising overhead—such as insurance, bond premiums and small tools—should be precisely specified.

Where a change causes an increase in the amount of time required to complete the project, the contractor is entitled to a commensurate extension of the project schedule. It is prudent to include a contract provision stipulating that extensions are available only to the extent that (1) the work affected by the change is on the critical path at the time of the change, and (2) completion of the *entire* project is thereby delayed or extended.

## Differing Site Conditions

One of the most frequent performance problems encountered on construction projects of any nature occurs when the contractor encounters unexpected site conditions. Absent a risk allocating provision for such unforeseen conditions, prospective contractors will, quite justifiably, increase the amount of their bids to cover possible costs associated with the contingency of encountering such conditions. Experienced construction owners will frequently attempt to allocate this risk by including a Differing Site Conditions clause in the contract and providing data on subsurface conditions to prospective contractors.[6]

Under a conventional Differing Site Conditions clause, a contractor can generally recover additional costs incurred due to unforeseen site conditions which materially differ from those shown in the contract documents, such as unexpected rock or water. Recovery is also possible where actual conditions are of an unusual nature, differing materially from those ordinarily encountered on a project like that being constructed.

In the context of a design-build arrangement, the risk of differing site conditions creates an interesting dilemma. It is often the responsibility of the design professional to recommend and conduct a pre-bid site and subsurface investigation. Consequently, if a design-build contract incorporates the conventional differing site conditions concept, the design-builder may benefit by conducting an inadequate investigation.

Parties to design-build contracts for construction of energy projects may want to resolve this dilemma by negotiating a contract provision specifying an economical and prudent site investigation program to be undertaken by the design-builder. Thereafter, if actual conditions materially differ from those revealed by the design-builder's investigation, the design-builder would be entitled to an equitable adjustment for additional costs incurred.[7]

Under this proposed arrangement the owner can avoid paying a windfall in the form of a contingency amount included in the design-builder's bid to protect against the possibility of unforeseen conditions which may never materialize. The contractor's risk is reduced since it can expect additional compensation if unforeseen conditions are experienced and, as a result, its overall price should be lower. As an alternative, owners should consider paying the design-builder to perform a detailed pre-design site investigation prior to contracting for design-build services.[8] Owners who insist upon requiring a design-builder to fully assume the risk of unforeseen conditions should expect to pay a sizable premium.

Another issue that should be addressed as a potential risk is the presence of contaminated soil or waste generated from the host facility—particularly if the host is a refinery or user of hazardous materials. Absent a specific contract agreement, there is a question as to whether the presence of waste material would be a differing site condition so as to justify relief to the contractor. Because of the potential magnitude of dollars associated with a cleanup plan, the parties should agree who, as between the owner and contractor, will bear the risks of this cleanup.

**Force Majeure**

Virtually all modern construction contracts contain provisions which excuse the contractor's failure to perform where the failure is due to causes beyond its reasonable control. These are known as *force majeure* provisions. These provisions specify the events that are deemed to be beyond the control of the contractor, which will justify a time extension to the scheduled date of plant completion. Typical *force majeure* events may include floods, civil, governmental or military authority, insurrection, riot, embargoes, strikes, acts of God or the public enemy or unusually severe weather.

Some specific performance problems are peculiar to power generation projects and may impact the *force majeure* clause. For example:

1.  *Approvals or permits from regulatory and environmental agencies.* In order to avoid disputes over whether the delays to this process are excusable, the contracting parties should define which, if any, regulatory delays will constitute *force majeure* events. It is also important to determine whether such delays are compensable, or whether the contractor is simply entitled to a time extension.

2.  *Technical problems at the host facility.* This can be a critical issue, since work may be stopped for reasons beyond the control of either the owner or the contractor.

3.  *Equipment delivery delays.* Many owners on power generation projects specify that certain major items of equipment, such as a turbine or boiler, be supplied by a designated manufacturer. These major manufacturers typically use their own standard form contract provisions which broadly define *force majeure*. In these cases, the parties should consider whether to incorporate a separate *force majeure* provision for the work and equipment supplied by these major manufacturers.

### Liquidated Damages

When construction projects are not completed on time because of the contractor's unexcused delays, it is frequently difficult to calculate the amount of damage to the owner. Furthermore, even if actual damages can be calculated, the calculation is often the subject of major disputes between the owner and contractor. Therefore, to avoid the risk of being unable to prove up actual damages, it is prudent for an owner to insist upon a liquidated damages clause that stipulates the amount of damages for each day of delay to project completion.

There are several issues associated with these types of clauses of which an owner should be aware. First, courts will require that the liquidated damages be a reasonable forecast of damages to be actually incurred by the owner and that they not be a penalty. A well-drafted liquidated damages clause should expressly acknowledge that delays will result in owner damages which are difficult to determine and that the parties agree to the stated amount as liquidated damages and not a penalty.[9]

It is also advisable for owners to remember that liquidated damages are not to be a substitute for the damages incurred in completing a contract where the contractor has defaulted or abandoned the contract. In these cases, actual excess completion costs may be recovered *in addition to* liquidated damages. The contract would expressly delineate such rights to the owner.

### Performance Guarantees

One of the unique features of a power generation construction contract is that the owner generally seeks, and the contractor is willing to provide, performance guarantees for certain aspects of the facility. These guarantees may relate to electrical or thermal output, noise emissions, air quality issues, fuel efficiencies, or myriad other aspects of the plant that are critical to achieving the financing or technical objectives of the owner.

There are several risks that an owner should remember when insisting on performance guarantees from its contractor:

1.  *Ensure that a sound mechanism exists for determining whether the contractor has achieved the performance levels required.* This is typically done by specifying detailed testing procedures. Among the items that should be addressed are the protocol for the test (which the owner should have the right to approve), acceptable tolerances in the test results, the duration of the tests and the remedies available in the event the test is not successfully completed.

2.  *Owners should recognize that the performance guarantee will be no broader than that specified in the contract.* For example, an issue often arises as to whether the contractor will guarantee the performance of systems and subsystems in the facility—since a system may not be working properly (i.e., running in excess of capacity and subject to premature burnout) with the plant producing as required. Unless specifically addressed there will likely be no guarantee for a malfunctioning system other than typical warranty requirements.

3.  *Level of guarantee vs. cost of the guarantee.* Contractors will extract a price for the guarantee being requested by the owner. Therefore, an

owner should establish guarantees that are consistent with the overall operational objectives of the plant.

4.   *Using "buy-down" amounts to achieve your objectives.* "Buy-down" amounts are similar to traditional liquidated damages in that they attempt to compensate the owner for failures by the contractor to achieve output performance guarantees. By paying the "buy-down," the contractor is typically relieved of further responsibility for schedule liquidated damages and for continuing efforts to successfully complete the performance tests. Owners should find this to be an acceptable alternative if the amount of the "buy-down" bears a reasonable relation to the diminished capacity.

### Performance Bonds and Guarantees

One of the risks that an owner faces is with the inability of the contractor to meet its obligations under the contract. A way to deal with this risk is to require that the contractor furnish payment or performance bonds. Performance bonds secure satisfactory performance of the contract and completion of the construction project. The surety is bound to the owner, to the extent of the amount of the bond, for the contractor's obligation to finish the project on time and in a workmanlike manner. Payment bonds are written for the benefit of subcontractors, and ensure that the subcontractors will be paid for their services on the project. This is especially helpful in states with liberal mechanic's lien statutes.

Some owners are willing to waive bonds in favor of a guarantee from a third-party that the contract will be completed in accordance with its terms. From an owner's perspective, this guarantee should be sufficient to protect it from the financial consequences of a contractor default.

# EFFECTIVE PROJECT MANAGEMENT

Many risks that an owner faces can be avoided if those responsible for contract administration on behalf of the owner follow some basic rules of sound project management.

### Scheduling

The owner of a power generation facility should have a strong understanding of the scheduling methods that will be used by the contractor to undertake the program for construction and complete the project in a timely manner. Sophisticated methods of scheduling, such as the critical path method (CPM), are widely used to plan activities and forecast critical delays. When properly used, the project schedule is a management tool that enables the owner to obtain advance warning of situations that may threaten the profitability of the project.[10]

A question that is frequently asked in the scheduling area is whether or not the owner should approve the schedule of the contractor. There are compelling reasons as to why an owner *should not* approve the schedule. Several courts have held that if the parties agree that the CPM schedule is a reasonable plan for performing the work, the schedule is presumed correct. Because the owner has no control over construction means, methods, manloading or economic restraints, it is virtually impossible for an owner to be in a position to vouch as to the validity of the schedule. Nor would an owner want to—since its true concern should be the date that completion milestones are actually met, not how the contractor intends to achieve such milestones.

It is, however, critical for the owner to evaluate the schedule and determine whether it is being required to perform its services in a manner consistent with its understanding of the business terms of the contract. For example, the schedule could call for a turnaround time on approvals of submittals in a shorter time than is reasonable, placing the owner in a position of delaying the contractor. Moreover, care should be taken to determine if the dates for bringing fuel to the site (which is typically the owner's responsibility) is consistent with the other agreements the owner has entered into on the project.

The project schedule should also be used as an analytical device for claim recognition, preparation, and proof.[11] This will enable the owner to have objective data as to whether delays are excusable under the contract, and will allow the owner to determine in advance if the project will be delayed by proposed changes.

Special care must be taken by the owner who has assumed the risk of contracting with various parties—such as equipment vendors and an erection contractor—to complete the construction of the project. In these multi-prime contracting projects, the owner is generally considered to assume duties analogous to those of the normal prime contractor to schedule and coordinate the work.[12] The owner's responsibility in this regard includes taking steps to require timely completion of one prime's work to prevent delay or interference to another prime, as well as scheduling work in a way that will allow each prime to perform economically where their respective work physically interrelates with that of other prime.

### Documentation

Another important administration tool for avoiding risks on the construction project is the creation, transmittal, control, and retention of project records and documents. During construction, a construction owner should establish and maintain systems that (a) identify the type, quality, frequency, and distribution of the records to be handled, (b) ensure that disciplined standards of documentation and proof are maintained, and (c) ensure that records are being preserved daily on every element of project administration and performance to permit a third party to reconstruct the project from the files, if necessary.

The records maintained on the project should include general correspondence, schedules and updates, minutes of job or coordination meetings, daily and weekly reports, memoranda for record, job diaries, progress photographs or videos, test and inspection reports, and weather data. Also important are accurate records on change orders, shop drawings, and payment applications. Job records are the principal source of evidence for resolving disputes and minimizing the potential for claims.

In the event delay or disruption is being claimed by the contractor, the owner's job-site staff should attempt to prepare internally an analysis of the causes and effects of the problems. If records are well-kept, they will assist in the practice of preventive contract administration. You will not only be well informed about the contract's performance history, but will also be better able to anticipate problems before they arise or become critical.

# INNOVATIVE DISPUTE RESOLUTION TECHNIQUES

Despite all the precautions taken by the parties in negotiating the scope of work and detailed project requirements, disputes can and do arise. Because litigation and arbitration are generally in no party's best interest, an owner should consider alternative dispute resolution techniques that are designed to handle the dispute quickly and cost-effectively.

## Mediation

Mediation is a process that is being increasingly used in construction disputes.[14] The mediator acts as a facilitator but generally has no authority to render binding decisions. In order for a mediation to succeed, the process demands full participation of all parties, represented by individuals with settlement authority. In addition, it is critical for the parties to have an unbiased decision maker involved to promote the settlement of the case.

## Mini-Trial

The concept of a mini-trial has also gained increasing acceptance in complex litigation. The name is somewhat of a misnomer in that a mini-trial is not a trial at all. Instead it is a structured settlement procedure and is usually voluntary and non-binding. In essence, the parties present the salient elements of their claim during a limited period of time. A summary of the evidence and relevant law is then presented and the "decision makers," consisting of a principal of each disputing party and a neutral advisor, attempt to render an opinion.

## Project Dispute Board

Another method of resolving disputes on the construction of power generation facilities is to establish, at the outset of a project, an individual or team of individuals capable of analyzing the technical and legal merits of project disputes. This concept has worked well on large construction projects, particularly if the Dispute Board meets regularly and is apprised of the project's progress.

# CONCLUSION

Owners and developers of power generation facilities should recognize that by carefully reviewing the project before construction starts, risks will be identified and contingencies made for dealing with such risks during contract performance.

## References

1. This article has been adapted from Loulakis, Gilmore and Hurlbut, "Contracting for the Construction of Power Generation Facilities," Construction Briefings No. 89-5, *Federal Publications* (1989).

2. *See generally,* Loulakis, Thompson and West, "Managing Construction Risks—The Owner's Perspective," Construction Briefings No. 91-5, *Federal Publications* (1991).

3. *See generally,* Loulakis and Love, "Exploring the Design-Build Contract," Construction Briefings No. 86-13, *Federal Publications* (1986).

4. *Id.*

5. *See e.g.* AIA Document A201, Article 7 (1987 ed.); EJCDC Document 1910-8, Article 10 (1983 ed.); FAR 52.243-4 (1987).

6. *See generally,* Currie, Abernathy and Chambers, "Changed Conditions," Construction Briefings No. 84-12 *Federal Publications* (1984).

7. *See generally,* Loulakis & Love, "Exploring the Design-Build Contract," Construction Briefings No. 86-13, *Federal Publications* (1986).

8. Id.

9. *Id.*; Louis Lyster General Contractor, Inc. v. City of Las Vegas, 489 P. 2d 646 (N.M. 1971).

10. *See generally,* Wickwire, Hurlbut & Shapiro, "Rights & Obligations in Scheduling," Construction Briefings No. 88-13, *Federal Publications* (1988).

11. *See generally,* Smith & Love, "Scheduling & Proof of Claims," Construction Briefings No. 82-6, *Federal Publications* (1982).

12. *See Stephenson Associates Inc.,* GSBCA No. 6573, 86-3 BCA ¶ 19,071.

13. *See generally, Mediation—Its Forms and Functions,* 44 So. Cal. L. Rev. (1971).

*Chapter 7*

# Cogeneration Start-ups in Retrospect

# Chapter 7

# Cogeneration Start-ups In Retrospect

*Freeman Kirby*
*Destec Energy Inc.*

A critical period in the development of a cogeneration plant, as with the development of all high-tech process plants, is that transition from concept and design/construct to full operation. This period is called start-up. This period represents a vital point in development and economic viability for the company. Any methods or practices which optimize and otherwise shorten this period of time is like finding a gold nugget in your own back yard. The collated experiences of several such start-ups are presented here.

Power Systems Engineering, Inc. (PSE) is a wholly owned subsidiary of Destec Energy Corporation and is located in Houston, Texas. The comments which follow are based on procedures worked out during our experiences in completing more than 20 cogeneration projects which vary in power generating capacity from 5 megawatts to over 465 megawatts.

Each of PSE's plant start-ups can be described in four basic phases.

1. Pre-start-up activities during design/construction phase
2. Actual start-up
3. Initial operation
4. Operation growth period

Each cogeneration plant start-up involved two basic teams. One is the Operating Team; the other is the Start-up Team. The start-up team's responsibility consists primarily of inspection of equipment, checkout of systems, plant start-up and plant performance testing.

## PRE-START-UP

The pre-start-up activities of the cogeneration plant begins with the assignment of an O&M Manager to the Project Team. The O&M Manager in conjunction with the start-up team provides a project overview consisting of the following:

1.  Process design review for the operability and maintainability of the cogeneration plant as well as integrity reviews of process flow diagrams and engineers' specifications to insure adequate process design.

2.  Assignment of the start-up team.

3.  Assignment of the operating and maintenance staff for the continuous operation of the plant.

4.  Preparation of the O&M manual.

5.  Preparation of the mechanical/electrical/instrument check sheets.

6.  Preparation of a detailed start-up schedule.

7.  Preparation of special start-up procedures.

8.  Preparation of a lubrication schedule.

## ACTUAL START-UP OF COGENERATION PLANT

When sufficient pieces of equipment are completed by the engineering construction contractor, the start-up team initiates the start-up activities. These activities begin with the first major systems that are needed for plant operation. An example would be an instrument air system. This system is checked out and placed in service to provide plant air to all instrument air systems. During this phase the following functions are completed:

1.  Loop checks are tested by the contractor and accepted by the start-up crew.

2.  Electrical functional test.

3.  Pre-mechanical checks are completed.

4.  Lubrication schedule completed.

5.  Motor alignments are checked and motors are bumped for proper rotation.

6.  Boilers are hydrostatically tested and chemically cleaned.

7.  All piping is hydrostatically tested and flushed.

## INITIAL OPERATION OF PLANT

During this phase, each of the major systems is started in the normal operating mode and checked for normal operations as a system. Baseline data are collected for the initial phase. Alarms and trips of major functions and loops are verified. The following are some additional activities that need to take place during the initial operation period.

1.  Collect baseline operation data for each major system.

2.  Verify that major equipment trips systems and protective systems are functional.

3.  Perform pre-parallel tasks for the associated utility company.

4.  Perform operational checks required by each supplier or vendor.

5.  Run system load check taking baseline data at each load increment defined in the test procedure.

6.  Run a plant performance test.

7.  Turn over the operation of the plant to the normal operating and maintenance staff.

# OPERATIONAL
# GROWTH PERIOD

The foregoing basically outlines the steps PSE went through for the start-up of each cogeneration plant. Generally, these procedures were applied to each of the plants. The construction schedule for these plants varied from a low of 9 months to a high of 12 months.

After the initial start-up of the plant and during the beginning operation, there was a period when most of the design problems, construction problems, operating problems were being identified and corrected. These problems caused delays in operations, reduced availability and revenue. This growth period in the cogeneration plant lasted from 9 to 12 months, depending upon various circumstances.

Minimizing the operational problems reduces the amount of down time of the cogeneration plant and enhances the revenue or power produced in the cogeneration plant. With an aggressive schedule in mind, and looking at the start-up of these plants in retrospect, the following is a list of ideas which when successfully applied, will reduce the start-up schedule and enhance the initial operation of the plant, thus reducing adverse operations during the initial growth period.

1.  **Design Review**
    During the design review process an extensive effort must be made to ensure that contract vendors' designs both mesh and coincide with the plant's process design. The operability and maintainability of the plant must be adequately weighed during the review of each vendor package. Process control features should not be relinquished during the economic evaluation.

2.  **Inspection of Vendor-Supplied Systems**
    The vendor supply systems should be inspected at two stages to ensure the proper operation and ease of start-up. To insure that the equipment meets or exceeds the design specification, inspections should commence during the manufacturing process and prior to shipping from manufacturing facilities. The second inspection should be in the field during the pre-operational checks where each piece of equipment is checked for the proper process function.

3.  **Inspection of Contractor-Supplied Equipment and Systems**
    During the construction phase, each system and each contractor-supplied piece of equipment must be inspected. To ensure that each system is installed per the engineers' drawings and specifications, this inspection must be accomplished in conjunction with construction activity to prevent costly rework and time delays.
    The emphasis on this inspection is directed primarily to an aggressive construction schedule. During the construction phase, the inspection must take place at the time each system is installed. With an aggressive schedule there is minimal time for removal of a system and reinstallation because the contractor did not follow the engineers' specification or the standard practices of construction. Inspection must be accomplished as these systems are installed or the rework will take place during the growth period of operation.

4.  **Performance Test**
    During the plant performance test, operational performance is tested for a complete plant in a pre-specified period of time. In addition to the normal performance test requirements, each major piece of equipment should be tested to ensure performance at the baseload condition. Additional data should be taken at different load requirements to ensure that each individual piece of equipment will perform per the engineers' specifications once in final operation.

5.  **Operating Staff**
    Due to the complexity of the cogeneration plant, the operating staff must be technically trained and established prior to initial operation. The operating staff must be formed well before the start-up date to ensure adequate time for training. The operating staff should be at the site 90 days prior to the expected start-up. During this period the operator training and operator assessment program must be implemented. Some of the operating technicians will be assigned to the start-up crew as the start-up progresses.

6.  **Spare Parts**
    Because of the complexity and high technological requirements of the plant, the spare parts should be divided into two groups:
    1.  Start-up spares
    2.  Long-time operating spares

The start-up spares must be ordered and expedited for delivery prior to the initial commissioning date. Some of the commissioning spare parts, especially cards for electronic devices, may not be purchased as commissioning spares but would be identified with the vendor as spares to be made available by the vendor in the event a failure occurs during initial operation or during the warranty period.

## SUMMARY

For future cogeneration plants to ensure a smooth commissioning period and a reduction of the initial operation growth period in an effort to maximize the revenue, the operating group must be on board and trained prior to the commissioning period. The planning of the commissioning period should be accomplished during the design stage. By placing emphases on these three areas, future cogeneration commissioning and initial start-up periods will go smoothly and ultimately improve the cogeneration plant's operational quality.

*Chapter 8*

# Reliability of Natural Gas Cogeneration Systems

*Chapter 8*

# Reliability of Natural Gas Cogeneration Systems

*Gas Research Institute*

C ogeneration systems fueled by natural gas exceed the reliability of most central station power generating units, according to a study conducted by ARINC Corporation for Gas Research Institute (GRI).

In the study, researchers obtained operating data from 122 natural gas cogeneration units nationwide representing 2,200 megawatts (MW) of capacity and nearly 2 million hours of operating time at 37 facilities. Units were grouped into categories reflecting size (from 60 kilowatts to 100 MW), type of system (gas engine or gas turbine technology), use of emission controls, and type of thermal application.

Various types and sizes of gas systems reported average availability factors ranging from 90.0 to 95.8 versus a weighted average of 85.9 percent for fossil-fuel steam, nuclear, and gas-turbine-based central station power generating units. Comparisons are based on study data and data reported by the North American Electric Reliability Council for utility power plants.

Gas cogeneration can improve utility operations because as a group the relatively small, dispersed cogeneration units are more reliable than one or more large central station units of similar capacity.

Cogeneration developers favor natural gas cogeneration systems by a wide margin because of their high efficiency, environmental compliance benefits, low costs, and short construction lead times (Figure 8-1). Such systems can be sized from a few kilowatts to hundreds of megawatts depending on electrical and thermal energy requirements.

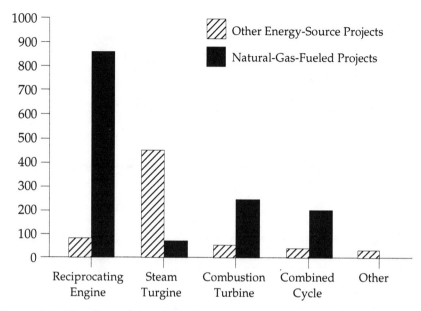

**Figure 8-1. Number of Cogeneration Projects by Type of Technology**

The reliability of gas cogeneration systems is a concern to electric utilities, which must maintain adequate generating capacity to serve all customers, including cogeneration users who occasionally require backup power supply. Reliability is also very important to the cogeneration system's owner or operator. Systems that operate reliably provide maximum economic benefits to the user. Conversely, unexpected system failures have a strong negative impact on economic viability because of the high cost of emergency repairs and premium rates for backup power (Figure 8-2).

To quantify the operational reliability of various gas-fueled cogeneration systems, GRI initiated the development of a comprehensive data base on gas cogeneration reliability. The results provide greater detail and more accuracy than any previous studies. Useful in evaluating both existing and planned cogeneration projects, the results should also help cogeneration operators, manufacturers, system packagers, engineering consultants, and others to identify cost-effective improvements that further enhance operational reliability and economic performance.

In the study, researchers obtained and screened operating and maintenance data from 122 gas cogeneration units representing 2,200

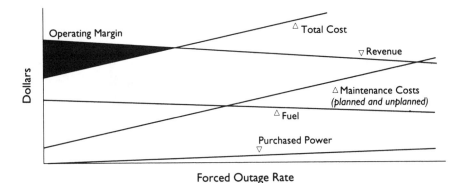

Figure 8-2. Impact of Outages on Operating Cost

MW of capacity and nearly 2 million hours of service time at 37 host facilities. The evaluation focused on systems driven by gas reciprocating engines and turbines, which represent the majority of all operating and planned cogeneration installations. Systems were grouped in six categories by size. Reciprocating engine systems ranged from 60-kW autoderivative engine systems to 6.5-MW dual-fuel systems; gas turbine systems ranged in size from 1.1 MW to 104 MW. Sixteen technology subcategories were also defined to detail prime mover characteristics and ancillary equipment.

When compared with operational data reported by the North American Electric Reliability Council for large central-station power plants, each group of gas cogenerators demonstrated better averages for key reliability measures (Table 8-1).

One important reliability measure is the availability factor, which reflects the total time the system is available for operation. Reported data for gas-fueled cogenerators show average availability factors in the range of 90.0 percent to 95.8 percent, compared with a weighted average of 85.9 percent for central-station plants. Another key measure is the forced outage rate, which indicates the time that the system is not available for operation as expected. The study found a forced outage rate of only 2.1 percent to 6.1 percent across the six categories examined, compared with 24. 7 percent for utility plants. These results indicate that gas cogeneration systems are much more reliable than commonly thought. Moreover, gas cogeneration can improve utility operations because as a group the relatively small, dispersed cogeneration units are more reliable than one or more large central-station units of similar capacity.

**Table 1. Reliability of Natural-Gas-Fueled Cogeneration Systems**

| Operational Reliability Measure[b] | Reciprocating Engine | | | Gas Turbine Engine | | | Electric Utility[a] |
|---|---|---|---|---|---|---|---|
| | Group 1 60 kW | Group 2 80-800kW | Group 3 >800 kW | Group 4 1-5 MW | Group 5 5-25 MW | Group 6 >25 MW | 1986-1990 |
| Availability Factor, % | 95.8 | 94.5 | 91.2 | 92.7 | 90.0 | 93.3 | 85.9 |
| Forced Outage Rate, % | 5.9 | 4.7 | 6.1 | 4.8 | 6.5 | 2.1 | 24.7 |
| Scheduled Outage Factor, % | 0.2 | 2.0 | 3.5 | 3.0 | 4.1 | 4.8 | 9.9 |
| Service Factor, % | 63.0 | 68.8 | 80.0 | 85.3 | 85.2 | 92.5 | 40.0 |

[a]*Average values are weighted by unit-years for fossil-boiler, nuclear, jet engine, gas turbine, and combined-cycle units from data reported in Generating Unit Statistics, 1986-1990, North American Electric Reliability Council/Generating Availability Data System.*

[b]*All figures are averaged. Operational reliability measures are consistent with American National Standards Institute/ Institute of Electrical and Electronics Engineers' Standard 762.*

For each of 9,500 recorded failure and outage events, researchers assigned standard cause codes consistent with the Institute of Electrical and Electronics Engineers' Standard 762. A specially created data base structure was developed to calculate statistics for several important operational reliability measures including the availability factor and forced outage rate. Because cogeneration units may operate in cyclic patterns, such measures provide a meaningful indication of a unit's ability to produce energy during its demand periods.

Detailed data on the performance of cogeneration subsystems identified component and subsystem improvements that could further enhance the reliability of gas systems (Figure 8-3, following page). For example, in Group 1 (autoderivative reciprocating engines from 60 to 75 kW), the engine and plant services subsystems together accounted for 49 percent of total unit forced outage hours. In particular, many failures involved high engine-coolant temperature resulting from inadequate system design and installation practices.

In gas turbine Groups 4 (1 to 5 MW) and 5 (5 to 25 MW), electronic control failures were the greatest contributor to forced outage events. Follow-on research is planned to expand the statistical basis of these findings and to address the need for subsystem and component improvements.

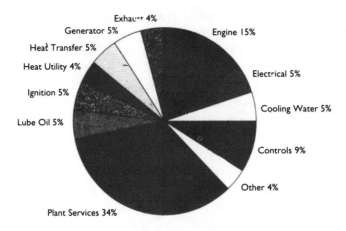

**Forced Outage Hours**
*(Total: 20,035 hr)*

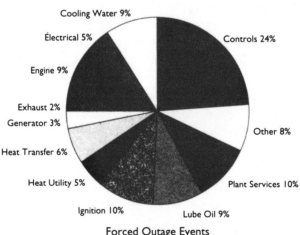

**Forced Outage Events**
*(Total: 618 Events)*

**Figure 8-3. Subsystems Contributing to Unreliability (Group 1)**

*Chapter 9*

# How Reliable are IPP/Cogen Projects?

*Chapter 9*

# How Reliable are IPP/Cogen Projects?

*Thomas E. Kalin, Donald O. Swenson*
*Black & Veatch*

The total electric generating capacity from new U.S. power plants constructed by nonutility (IPP/Cogen) power producers is now more than the new capacity from plants constructed by the regulated power companies. How reliable will these IPP/Cogen projects be? And the corollary to that question, how can these plants be made more reliable and profitable?

This chapter discusses one approach being taken to answer the first question, and discusses five criteria for the achievement of a reliable, profitable project.

## DATA TO ASSESS RELIABILITY

Regrettably, most people in the industry, if asked how reliable are IPP/Cogen units, would have to answer "No one knows." The reason for this is that unlike for the regulated utility generators, there is presently very little data available for analyzing and comparing the reliability of IPP/Cogen projects with that of utility owned projects.

The most complete and by far the largest source of reliability data for existing electric generating units is the North American Electric Reliability Council's Generating Availability Data System (NERC GADS). Over 90 percent of this country's electric utilities are included in the data base.

This data includes length of outage, duration, capacity lost, type of outage, and cause for each outage. The data also includes design information such as unit type and age, fuel type, steam generator and turbine manufacturers, numbers of certain major components, types of pollution control systems, etc.

Examples of how some of the information may be used are shown on Figures 9-1 through 9-5. From GADS data for utility units, for example, Figure 9-1 illustrates the distribution of average equivalent availabilities, net capacity factors, and forced outage rates for utility coal fired units smaller than 100 MW for a 5-year period (1982-1986).

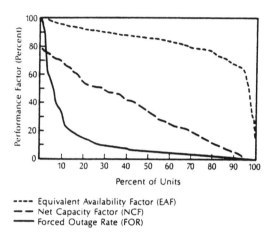

---- Equivalent Availability Factor (EAF)
— — Net Capacity Factor (NCF)
—— Forced Outage Rate (FOR)

**Figure 9-1. Distribution of Five-Year Average (1982-1986)
Performance Statistics for Coal Fueled Units Smaller than 100 MW**

A significant aspect of some of this information is that most of these small coal units are over 30 years old, as shown on Figure 9-2, and therefore the information is dated. Current data for new units, both utility and IPP/Cogen is drastically needed.

Reliability data for other types of units which are often employed in IPP/Cogen projects are shown on Figures 9-3, 9-4, and 9-5. These figures show the distributions for combustion turbines, combined cycle units, and multi-boiler, multi-turbine units. Overall availability averages for these unit types are shown in Table 9-1.

Fortunately, a system has recently been established for collecting data concerning the nonutility generating plants. This system, described in Data Reporting Instructions for Nonutility Generator Units, has also been prepared by the North American Electric Reliability Council.

What is unusual about this data collection is that it is being collected through the electric utilities that purchase the electricity generated from

### Table 9-1. Overall Reliability Average for Units Less Than 100 MW Forced

| Unit Type | Equivalent Availability percent | Net Capacity Factor percent | Forced Outage Rate percent |
|---|---|---|---|
| Coal Fueled Units | 81.8 | 36.9 | 7.3 |
| Combustion Turbines | 84.6 | 1.5 | 60.9 |
| Combined Cycle Units | 75.9 | 28.0 | 5.9 |
| Multi-boiler, Multi-turbine Units | 84.9 | 40.5 | 5.9 |

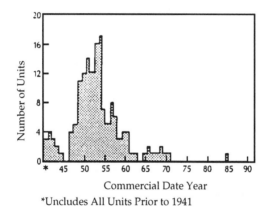

Commercial Date Year

*Uncludes All Units Prior to 1941

### Figure 9-2. Age Distribution (Commercial Date) for Coal Fueled Units Smaller Than 100 MW

the IPP/Cogen projects, then submitted to NERC GADS for processing. The forms used by utilities to collect design information and operating performance data are similar to those used for utility data, as seen in the GADS monthly operating data form for IPP/Cogen units, Figure 9-6.

In addition to the design and fuel information for each project, operating data requested includes hours for unit service, reserve shut-down, pumping, synchronous condensing, total available, planned outage, maintenance outage, extension of planned outage and maintenance outage hours, scheduled outage, forced outage, total unavailable, total unit derated, and average size of reduction. The collection of these types of information will, of course, not make the plants reliable, but data on such items as the above will help determine how reliable, and thus how profitable IPP/Cogen plants can be.

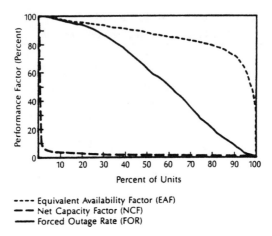

**Figure 9-3. Distribution of Five-Year Average (1982-1986) Performance Statistics for Combustion Turbine Units**

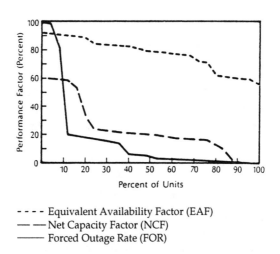

**Figure 9-4. Distribution of Five-Year Average (1982-1986) Performance Statistics for Combined Cycle Units**

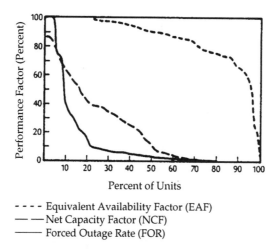

Percent of Units

- - - - Equivalent Availability Factor (EAF)
———— Net Capacity Factor (NCF)
——— Forced Outage Rate (FOR)

**Figure 9-5. Distribution of Five-Year Average (1982-1986)
Performance Statistics for Multi-Boiler, Multi-Turbine Units**

# CRITERIA FOR SUCCESSFUL PROJECTS

Until IPP/Cogen units can be shown to be reliable on the basis of actual design and performance data such as that being prepared by NERC, a number of criteria can be considered, which will help to make nonutility generating projects both reliable and profitable.

Reliability impacts IPP/Cogen projects in many ways. Among the areas affected are electric and thermal energy sales and income, potential penalties from contractual reliability clauses, and effects on the IPP/Cogen owner's profitability and capability to repay financial institutions.

Black & Veatch has performed reviews to determine technical viability and financial viability in project and financial assessment work for owners and financial institutions for over 50 IPP/Cogen projects during the past five years. Of the IPP/Cogen projects, designed, constructed, or placed into operation in that time, the successful ones have met the following criteria:

- **they have been well-planned, coordinated, and managed;**
- **they have used conventional technology and conventional fuels;**
- **and they have been properly funded.**

**A. UNIT IDENTIFICATION**

Data Code

| 7 | 5 |
| 1 | 2 |

Utility Code

3          5

Unit Code

6          8

Year

9          12

Month

13    14

Revision Code

15

**B. UNIT GENERATION INFORMATION**

Net Maximum Capacity (MW)

16          21

Net Dependable Capacity (MW)

22          27

Net Actual Generation (MWh)

28          34

**C. UNIT STARTING PERFORMANCE**

Typical Unit Loading Characteristics

35

Description (only if 6 was used in column 35)

36          60

Attempted Unit Starts

61    63

Actual Unit Starts

64    66

**D. UNIT TIME INFORMATION**

Service Hours

67          71

+ Reserve Shutdown Hours

72          76

+ Pumping Hours

77          91

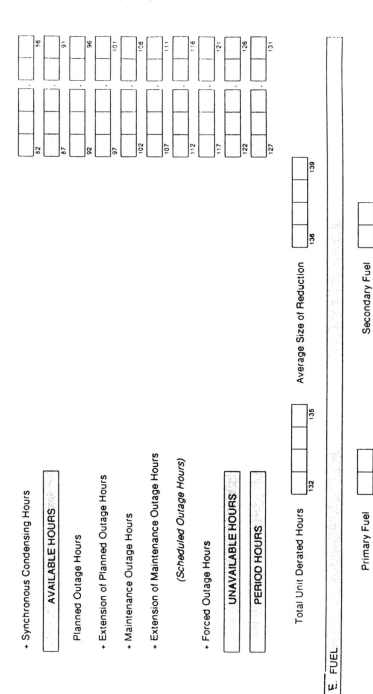

Figure 9-6. NUG Performance Report Format

Those which have not fulfilled one or more of these criteria have generally not been reliable nor profitable. The following paragraphs describe these criteria and provide examples showing the effects of failure to meet these criteria.

## Well Planned and Coordinated

Good planning and coordination are needed continuously from initiation through completion of the project, and the lack of planning and coordination quickly becomes evident and has lasting effects. On many projects, the lack of start-to-finish planning with no single entity responsible for execution has reduced both project effort and success. At one project, for example, the lack of planning for adequate public support resulted in new local regulatory restrictions being placed on the project, which restrict both operation and fuel type.

At another project, lack of coordination and failure to follow proper permit procedures delayed the receipt of the required permits, delaying the project more than 18 months. Construction of this project has still not been started.

At a third project, lack of planning resulted in the failure to establish a firm supply of the primary design fuel; therefore, the more expensive backup fuel has become the primary fuel. This has increased operating expenses, reducing the cash available for financing long-term debt payment.

## Well Managed

Being well managed means, among other things, that project management stay informed and responsive on all aspects of the project, delegating where appropriate, and personally following up where appropriate, both in a timely manner. Not being well managed has caused more than one IPP/Cogen project to be unreliable or unprofitable.

One example of this was a project at which the operations and maintenance contractor did not fulfill his obligations, leading to low availability and the lack of proper maintenance. After two years that contractor was finally replaced, and the project has had to use valuable time and funds to repair and replace worn out and poorly maintained equipment.

## Conventional Technology

The old saw, "Be not the first by whom the new is tried, nor be the last to lay the old aside," may apply here. However, conventional technology does not necessarily have to mean only the basic fossil fueled applications. Conventional technology can be state-of-the-art, and it can relate to any system from fuel handling through electric and steam transmission.

The reason for using conventional technology is, of course, that not all potential problems can be anticipated with a new technology, leaving the nonutility generator open to construction delays and unanticipated breakdowns and maintenance. The first circulating combustion fluidized bed (CFB) boilers, had for example, various problems, including fuel feed and tube erosion complications. However, in the second and third generation CFB boilers, these problems have been solved and the technology has shown much better success in operation.

Nonconventional technology has also been used on coal handling systems, ash and solid waste handling systems, piping and electrical systems, and water treatment and chemical cleaning systems. Each of these nonconventional approaches has caused extensive maintenance time, lowering the reliability and profitability of the project.

## Conventional Fuels

Conventional fuels can be characterized and purchased fairly reliably, and most components can be relatively assured by design to burn the conventional fuels. This is not always the case when a project uses nonconventional fuels, and their use has been a cause of problems at a number of IPP/Cogen projects.

These projects have used nonconventional fuels such as waste bituminous and anthracite coals, tires, biomass, agricultural waste, and peat, and each has employed unproven technology to handle and burn these fuels. The penalty has been an expensive learning curve in extensive maintenance and design upgrade to improve the performance of these initially unreliable, unprofitable projects.

## Properly Funded

Sufficient funding is absolutely essential to complete construction, train operating and maintenance staff, provide adequate spare parts at the project site, and secure enough fuel on site for continuous operation of the project.

Without proper funding, projects have not been able to finalize construction and have had to restrict maintenance. Because of the reduced routine maintenance, more forced outages have occurred, again reducing reliability and profitability.

# TRENDS OF IPP/COGEN UNITS

Several trends have been identified for IPP/Cogen projects, and these trends will likely be documented as the NERC GADS data are collected. A readily evident trend is that larger units that are the size of conventional utility generating units are expected to be built by the IPP/Cogen owners.

Design of future IPP/Cogen projects will be changed and improved in several areas. Improved combustion gas turbines firing distillate oil will be a future approach, with development of improved distillate oil fired turbine combustors. Designers will more frequently employ steam injection into combustion turbines to augment electric power output when process steam demand decreases. They will utilize more heat recovery steam generators which can be independently fully fired, including the addition of a separate forced draft fan system.

Waste-to-energy projects will likely develop technological capabilities to optimize the system. The capability to remove noncombustibles like glass and aluminum cans from the refuse stream and also to remove high-moisture components like grass clippings and leaves will lead to new innovations for improved waste-to-energy projects.

Rotating cylindrical combustors may be used in larger waste-to-energy projects. The recent trend toward refuse derived fuel (RDF) projects illustrates how recycling has affected the industry.

Fluidized bed combustion (FBC) is also being considered for processing waste. The use of a 50 MW fluidized bed plant to burn a combination of wood waste, coal, and RDF is now in start-up.

Coal gasification/combined cycle technology will be an alternative for coal fired projects. The nation's first large-scale coal gasification/combined cycle project is being converted to gasify a mixture of 25 percent municipal sewage sludge and 75 percent coal.

# SUMMARY

IPP/Cogen projects will continue to supply half or more of the new electric capacity in the future. For these projects to be reliable and profitable, meeting the five criteria for successful projects which have been discussed in this paper is still necessary. Ensuring that these five criteria exist on your project enhances the potential for success of the project.

The IPP/Cogen industry must commit itself to participation in the new generating availability data system established by the North American Electric Reliability Council.*

The measurement of that success is dependent upon the development of a data base, which must be supported by the IPP/Cogen industry. Without that participation, it will be very difficult to answer the question, "How reliable are IPP/Cogen units?"

## Bibliography

Data Reporting Instructions for Non-Utility Generator Units, North American Electric Reliability Council, Generating Availability Data System, Princeton, New Jersey, January 1991.

Kalin, Thomas E., Kevin D. Jennison, and Donald O. Swenson, "The Role of Reliability in Independent Power Projects," presented at Cogeneration & Independent Power Market Conference, New Orleans, Louisiana, April 2-3, 1990.

Makansi, Jason, "New CFB Design Emerges in Pennsylvania Culm Fields," *Power*, July 1990.

Makansi, Jason, "Fluidized-Bed Boilers," *Power* March 1991.

"Non-Utility Use of Fluidized Bed Boilers: A Growing Technology," *Power Engineering*, August 1990.

"Record 224 IPPs Came On Line Last Year, Study Shows," *Cogeneration*, February March 1991.

Renault, J.P., "Availability of Combined Cycle Power Plants," GEC ALSTHOM Technical Review No. 1, 1990.

Smith, Douglas)., "Regulated Utilities No Longer the Leader in Building New Capacity," *Power Engineering*, March 1991.

Syngle, D.V., and B.T. Sinn, "CFB Boiler Fires Waste Coal, Achieves High Availability," *Power*, April 1991.

"Trends and Technology Update," *Power*, October 1990.

---

*Information on reporting of data for nonutility generators is available from NERC GADS and can be obtained from the following address:
North American Electric Reliability Council
Generation Availability Data System
101 College Road East
Princeton, New Jersey 08540-6601 Phone: (609) 452-8060

# Operation and Maintenance
# Costs of Cogeneration and IPP Plants

*Chapter 10*

# Operation and Maintenance Costs of Cogeneration And IPP Plants

*Rudy E. Theisen, P.E.*
*Destec Energy Inc.*

M any of the determinations that impact the Operation and Maintenance (O&M) costs of Cogen and IPP plants are made months and years before these plants are in actual routine operation. These early decisions are made during the five phases of a Cogen/IPP project preceding Routine Operation—(1) Project Development, (2) Engineering and Design, (3) Procurement, (4) Construction, and (5) Startup and all affect the long term O&M costs of a plant. This chapter explores the impact on O&M costs of these early determinations and suggests that an established Company experienced in all of these phases will produce the lowest O&M costs in Cogen or IPP plants.

## PROJECT DEVELOPMENT PHASE

This is the project phase that determines the economic feasibility of a project and tailors the Cogen or IPP plant to the general needs of a customer. After it has been determined that the proposed plant location is friendly with regard to environmental permitting, zoning, fuel/water/labor/site availabilities, and power and/or steam needs the general energy cycle and sizing of the major equipment such as gas turbines, boilers, steam turbine generator, cooling tower, is accomplished. A few of the Project Development criteria are listed below and related to routine operation O&M costs:

1)    Revenue And Return Upon Investment—Certainly the project must produce an acceptable revenue during the life of the project based not only on the usual financial considerations but also on the antici-pated economical, legal, and fuel-use attitudes and regulations. A change in an environmental regulation can affect O&M costs in the form of additional treatment facilities requiring chemicals, mainte-nance, and supervision.

2)    Required Reliability—What are the reliability needs of the power and/or steam customer? Is the Cogen plant supplying steam to a petrochemical plant where utilities reliability is extremely impor-tant to prevent such hazards as vapor releases, or to a batch process where the loss of utilities is of little consequence? Reliability in the form of redundant control systems, conservative designs, or extra operators can affect both initial capital and long term O&M costs. The same reliability needs would apply to an IPP plant with capac-ity incentives during peak periods.

3)    Dispatching—Will frequent dispatching of major equipment be re-quired? Which of the available major equipment in the market place handles cycling with the least amount of deterioration and maintenance requirements?

4)    Historical O&M Costs—What is the industrial record of O&M costs of the available major equipment that will fit the desired energy cycle and operating mode? What are the several year costs of the major spare parts that deteriorate and must be replaced on a peri-odic basis?

5)    General Plot Plan—Is there sufficient space available for equipment removal and laydown? Is contractor parking available for major equipment inspections?

6)    Effluent—Effluent disposal plans. Will a permit be required? Will long term effluent transportation be required?

# DESIGN/ENGINEERING PHASE

In this phase, specifications are developed, equipment selections made, operation and control philosophies determined, redundancies identified, major and expendable spares identified, materials of construction selected, O&M manuals prepared and most importantly the primary boiler feedwater water treating method and equipment is selected and sized based on historical quality records of the raw water supply. How do the preceding affect routine O&M costs?

1) Specifications—If specifications are inadequate, expensive field corrections could be required one time or continue for several months or years after plant startup. (Example: Trouble alarms not specified would have to be added in the field but possibly not before equipment damage occurs that causes circumferential stresses that translate into later troubles.)

2) Vendor Proposals—Proposals should be requested only from vendors who have demonstrated a favorable track record in the areas of quality control, accepting responsibility for defects, and generally being customer-oriented. It is not improbable for the O&M costs of a 100MW plant to vary by $50,000/year due to a single unresponsive major equipment vendor.

3) Equipment Selection—The equipment quality and reliability from the various vendors is cyclical depending upon the vendor's profit picture, experience level, and quality control. Equipment selections should be made from manufacturers that tend to be on the "high side" of the quality cycle. It would be embarrassing to purchase a troublesome and costly boiler feedwater pump to find that industry-wide the pumps had been in disfavor for several years.

4) Control Philosophies—Will this be a plant with: (a) a proven central distributive control system (DCS), or (b) a plant with the various equipment systems such as the gas compressor, demineralization equipment, and chemical feed systems having individual instrumentation packages located at the equipment site? If the DCS system is truly an effective total plant control, a 100-150MW plant

would probably require no more than two operators per shift or possibly one and one-half operators if sharing with a nearby plant is possible. Individual system control packages for the same plant would require a minimum of three operators to obtain the equivalent level of reliability as the DCS controlled plant.

5)   Redundancies—Redundant pumps, uninterruptible power supplies, electrical feeders, and backup gas turbine fuels all increase the reliability of a plant but also increase the preventive and corrective maintenance load. These redundancies have an upkeep cost. (Industry is beginning to think that in some cases the extra maintenance load and cost for redundancy is more than offset by the decreased stress and equipment costs caused by unnecessary full-load emergency trips.)

6)   Spare Parts—The availability of on-site spares for quickly changing out defective components eliminates the maintenance time and cost for "jury rigging," repairing when repairing is not economical and searching for available spares from other owners. Major spare parts should be included in the request for vendor proposals when the discount leverage is the greatest. Not doing this could result in paying list price one, two, or three years later.

7)   Materials of Construction—If the proper materials of construction are not selected, especially involving corrosion and high temperature, the O&M costs can be drastically affected. (*Example:* The materials of construction for a boiler feedwater pump handling 150 mmho water is not adequate for handling 10 mmho water. Selecting the improper material can create the need for periodic epoxy coating and weld overlaying causing higher maintenance costs for the life of the plant. *Another example:* A waste heat boiler duct liner should be selected based upon the gas turbine peak exhaust temperature, not the average, or lingering maintenance costs will occur.)

8)   Water Treating Method—Worldwide, this is probably the most neglected facet of power plant design and when poorly done can cause the overall plant O&M costs to increase by as much as 5% over a proper design. In the area of water treating, (a) reliability

should somewhat outweigh capital costs, (b) pioneering vendors should not be considered, and (c) equipment should be sized based upon a year or so of raw water analyses and a few per cent resin degradation. The controls and programming should be proven and blended with the plant DCS.

9)  Design Subtleties—Are all of the personnel involved, design, project, and operation in agreement with the methods for handling:
    a)  Waste effluent (remember, some of the effluent could be oily).
    b)  Cleanliness and pressure of the available fuel.
    c)  Noise control.
    d)  Obtaining building/construction/environmental permits.
    e)  Adjacent environment (chemical plants, dusty activities, ambient corrosivity).
    f)  Community makeup (bedroom or industrial?).

If agreement is not obtained during design, costly O&M "fixes" will occur during the normal operating years.

## PROCUREMENT

This is the phase of the project that encompasses purchasing, expediting, and receiving the equipment as defined in the design phase. What are some of the purchasing functions which, if performed improperly, could impact O&M costs?

1)  Purchase Order Preparation—If a purchase order (PO) does not cover all of the terms in the selected proposal or those negotiated at requisition meetings, the omissions could increase O&M costs. It is not uncommon for an equipment vendor to "throw in" a multiplying factor (discount) for future spare parts to seal a deal. It is also not uncommon for a vendor to provide the startup or first year parts in order to obtain an order. If the preceding are not precisely included in the PO, they are legally negated.

2)  PO Design Specifications—A copy of the latest version of the design, specification should be included with the PO. There is a known case where the second level technician training provisions

on a DCS had been omitted on early draft specifications but inserted in the final. However, the final version was not attached to the PO, and as a result, the second level training increased the O&M costs in year two and three of operations by about $15,000/ year.

3)    PO Approval—The POs should not be blindly signed by a project manager thinking, "surely everything is all right because so many have already reviewed the wording." Not so. Some POs have many pages but still should have a final review and comparison to the latest proposal, design specifications, and requisition meeting minutes.

4)    Expediting—Expediting, which includes scheduling, phone contacts, on-site inspections, witnessing shop performance tests, comparing shop fabrication to design specifications and collecting the final testing paperwork is the most important function of the procurement phase of a project. It is not difficult to imagine the annual O&M cost increase that could arise in one to five years if deficiencies existed in the following areas:

a)   Pump clearances.
b)   Transformer testing.
c)   Alloy heat treating.
d)   Turbine oil piping cleanliness.
e)   Pressure testing.
f)   Motor testing.
g)   Materials of construction.
h)   Mill tests.

An experienced expediter with formal check sheets is absolutely required in the seller's market of today to properly control the materials of construction, fabrication, and assembly of key equipment.

5)    Receiving—Receiving, checking, and especially properly storing purchased equipment to be constructed and the major spares for routine operation all too often receive little attention because it seems so simple, "Hire a clerk and check the stuff in." This causes trouble not only during construction but in later years of operation. A dry storage area arranged and cataloged for expensive valves,

motors, hydraulic operators, instruments, and transformers is required. Some electronic equipment may require air conditioned storage. A case is reported by a manufacturer where $50,000 of spare electronic control boards were received by a customer during construction and stored in a windowless construction shack. Six months after the completion of construction, one of the boards was required for a turbine repair. After several hours of searching, the corroded mass was located and had to be sent to the junk pile. The replacement required for the repair had to be air freighted, further increasing the O&M cost that had already been increased by purchasing another set of boards.

# CONSTRUCTION PHASE

The construction and startup phases are concurrent and deeply intertwined especially for the two months preceding plant startup. However, these phases will be handled separately since they impact routine O&M costs in different ways.

It is obvious that long term problems will develop and migrate to the routine operations phase as extra expense if a plant is poorly constructed due to such factors as unqualified constructors, inexperienced managers and inspectors, informal testing and acceptance procedures, not adhering to the construction prints, taking shortcuts, and thinly staffing the front line supervision.

During construction, the operating representative who is usually a startup specialist or the future plant superintendent who has been involved in all of the previous project phases, now becomes an extra set of eyes for the construction manager to aid in solving and preventing problems during construction rather than later and more expensively in the routine operation phase. Also as the O&M technicians are hired, usually three to five months before startup, they also become subordinate to the construction manager via the plant manager as inspectors, safety watches, and control checkers. A few of the specifics during construction that will prevent later O&M expense are:

1) Field Changes—Must be approved by the construction, design, and operating representatives.

2)   <u>Construction Contingency Allowance</u>—This phase converts the final single dimension plant design into a three-dimensional real life process. An allowance should exist in the construction budget for necessary field changes as the three-dimensional shape develops and uncovers:

a)   Head knockers.
b)   Inaccessible valves.
c)   Shortage of drains.
d)   Missing blind and disconnect facilities.
e)   Missing ladders.
f)   Maintenance obstructions.
g)   Missing walkways.
h)   Awkward manhole covers.

If the preceding are not corrected during construction, they become a routine operation revision expense.

3)   <u>Spare Parts Storage</u>—This was covered under the receiving section of PROCUREMENT, but again, the equipment and spares received during the construction period for routine operation must be cataloged, prepared, and stored in the proper atmosphere to prevent later increased O&M cost. Construction must provide the manpower and the storage to handle these requirements because many of the routine operation personnel have probably not yet been hired nor has a permanent warehouse been constructed.

## STARTUP PHASE

This is the project phase that begins one or two months before the performance test where commissioning, training (manufacturers and in-house), final control checkouts, development of startup checksheets, final safety preparations, and "steam blowing" occurs. In this phase a startup specialist or the plant manager assumes the role of startup coordinator and the construction manager's role subsides in support of the startup coordinator. The final modifications are made to the operating and maintenance manuals that were begun during the DESIGN PHASE. Three examples are listed below of how an effective STARTUP PHASE can favorably affect O&M costs.

1) Training—It is obvious that if maintenance technicians are not trained properly, poor workmanship, "redoes," and overtime to perform a high pressure pump overhaul, could easily increase an $8,000 job to $12,000. Also, poorly trained operators will make errors that cause damage to equipment not covered by manufacturer's warranties. A formal, several week, classroom and field training program is mandatory.

2) Commissioning—Each piece of electrical control and rotating equipment should have a commissioning checksheet that is completed and signed as satisfactory by the constructor, operator, and sometimes the manufacturer. An example of the checks to be performed on a 500 HP boiler feedwater pump might be:
   a) Coupling aligned and documented.
   b) Correct rotation.
   c) No bare wires, motor megged.
   d) Conduit covers secured.
   e) Pump filters in service.
   f) Lubrication okay (oil and pumps).
   g) Local and remote start/stop function.
   h) Automatic shutdown or start functions.
   i) Recirculation valve free and working.
   j) Proper drains installed.
   k) Seals checked.
   l) No unusual noise or overheating.
   m) Operate pumps with a mechanical technician present for 30-40 minutes.

It is obvious if the preceding checks are not performed, damage could occur such as a scarred bearing caused by a momentary oil starvation, that does not fail until one or two years following startup. Also, an untested malfunctioning low oil pressure automatic shutdown could cause pump damage two or three years after startup.

3) Final Safety Precautions—If the transformer sprinkler systems are not checked during startup, a fire could cause considerable O&M damage at a later date. Fire extinguishers not strategically located could result in an insignificant small fire becoming a damaging fire. Inoperative safety showers could result in a lost time accident and

resulting overtime, also, possibly a legal suit. Safety procedures and manuals should all be in place to prevent these and similar occurrences that can increase O&M costs.

4)    In-Service Checks—There are a series of activities that should take place as the equipment is actually placed in service.
      a)  Infrared scan of electrical busses, boiler ducts, and isophase ducts.
      b)  Steam, boiler feedwater, or fuel leak inspections.
      c)  Check for strange noises or odors from rotating or electrical gear.
      d)  Start and stop the gas and steam turbines. Do the auxiliary oil pumps operate as designed?
      e)  Major rotating equipment should have continual attendance at the site for one or two shifts following initial startup.

Expanding on 4a, it is easy to visualize what an insecure buss connection could cause in maintenance costs after startup if arcing and ionizing began.

# NORMAL OPERATION PHASE

This is the normal phase of operation that follows the startup phase and covers the day-to-day management, operation, and maintenance of the Cogen or IPP facility.

1)    Management—This activity covers proper staffing, communication, continued training, proactive safety, morale, salary administration, planning, leading, and controlling. It is obvious that deficiencies in any of these areas would cause an increase in O&M costs due to work "redoes," employee turnover, overtime, operating mistakes, and accidents. Proper management of a plant assures that the operation and maintenance is carried out with well-informed, trained, innovative, secure, and loyal O&M technicians.

2)    Operation—To prevent operational mistakes, the following are the minimum of the programs that should be in place:

a) Each new operator should be provided formal classroom and on-the-job training.
b) Graded examinations follow the training.
c) Each operator has a personalized training matrix on file so they know at the beginning of their employment what is expected.
d) The scheduled outages of major equipment is communicated well in advance so the operators have an opportunity to refresh themselves on shutdown procedures.
e) Startup check sheets exist for each piece of major equipment.
f) "What If" drills are practiced monthly.
g) Complete operating manuals and emergency procedures exist.
h) Safety procedures and isolation sketches exist for preparing equipment for entry and maintenance.
i) A formal risk vs. reward analysis is performed on any equipment or procedural revision.
j) Plant performance parameters are identified and continually monitored.

Again, it is obvious that the preceding decrease the chance for errors, oversights, and accidents that would increase O&M costs.

3)   Maintenance—To minimize the O&M costs of a Cogen or IPP plant the following maintenance programs should be in place:
a) Formal preventive maintenance (PM) schedules should be developed blending past experience and the manufacturer's recommendations. PM should be performed on overtime within a predetermined period if not completed by the due date.
b) Major equipment inspections should be performed per manufacturer's recommendations.
c) A predictive maintenance program exists consisting of at least visual, vibration, and boroscope inspections.
d) Handle minor corrective maintenance before it becomes major corrective maintenance.
e) Formal maintenance records are kept on each piece of major equipment.
f) An adequate stock of spare parts and consumables are kept on hand.
g) On very technically specific problems, manufacturer's representatives are utilized.

h) A computerized maintenance program should be implemented
which consists of:
(1) Preventive Maintenance.
(2) Corrective Maintenance.
(3) Stores (Inventory Control).
(4) Stores (Purchasing).
(5) Equipment History.

In conclusion, a full-service, established, and experienced company
that can handle Cogen and IPP project development, design, procurement, construction, startup, and routine operation is the company that
will provide the lowest long-term O&M costs.

*Chapter 11*

# Operation and Maintenance
# Considerations for Cogenerators

## Chapter 11

# Operation and Maintenance Considerations for Cogenerators

*Paul L. Multari*
*Mission Operation and Maintenance, Incorporated*

O
peration and maintenance is the key to financial success of any cogeneration project. Four criteria to use when evaluating an operation and maintenance company are described below. Cogeneration project owners, developers, and investors can use these four criteria to compare O&M services providers.

## CONTROL ENERGY PROJECT RISK THROUGH O&M

Operation and maintenance costs represent a relatively small percentage—around 2%—of a power plant budget. It's no wonder then that lenders, investors and developers often overlook O&M when evaluating a project.

The truth is, however, that O&M is one of the most significant factors contributing to the success of a power plant. A successful operation and maintenance program can make money for a project. An unsuccessful program can financially bankrupt an otherwise good project.

# WHY O&M IS IMPORTANT

The operation and maintenance organization is the only group which can provide revenues for the project once it is built. Typically, project financials are based on a given level of production. Since the O&M services provider is responsible for keeping the plant on line and producing to expected levels, it is the only group that can meet or exceed the capacity performance assumed in the project financials.

Simply stated, if the plant's not running, it's not making money. And the only way to keep a plant running is with quality operation and maintenance. Therefore, lenders and developers must ensure in the initial analysis that the O&M services provider can perform to the assumed levels. Otherwise, the plant will not achieve the hoped for financial goals.

Increased competition within the Independent Power Producer (IPP) industry makes evaluation of the O&M services provider a key factor in the comparison of competing projects. As more companies participate in the IPP industry and utilities offer less generous avoided cost contracts, good O&M can make the difference between being competitive or not.

The operating group can affect the plant's production on the margin and provide additional incremental revenues to the project. Thus, the operator's ability to meet or exceed pro-forma performance levels is critical.

# FOUR CRITERIA FOR EVALUATING O&M

There are four major criteria that O&M service companies should be evaluated on:
Experience,
Standards and Practices,
Quality of Personnel, and
Other Organizational Capabilities

### Experience
First, the O&M provider should have direct experience with IPP projects. Knowledge of IPP operations enables a company to operate and

maintain a project to maximize revenues achievable through the Power Purchase Agreements. This means that the operator must be able to take advantage of opportunities to benefit the project. For example, outages are commonly scheduled during "off-peak" hours when avoided cost payments are minimal.

But, during a recent cold snap, utility gas users in one service territory were temporarily curtailed, forcing them to switch from inexpensive gas to more costly fuel oil during the winter months. This resulted in an abnormally high avoided cost during the "off-peak" hours.

Our operating group took advantage of the increased avoided costs by rescheduling the outages until after the curtailment was lifted, thereby earning increased revenues for the project by staying on line. This is an example of how an O&M company can add value to a project.

A less dramatic example of how experience can affect a plant's profitability is the timing of off-line compressor washing. To clean the compressor on a gas turbine, the machine must be taken off line for several hours, reducing plant output and revenue.

However, after the procedure is completed, the machine will operate more efficiently and with increased output. By developing methodologies to optimize the timing of compressor washes, the revenue loss from taking the compressor off line for cleaning is optimally balanced by the increased revenue from compressor efficiency gained by the cleaning.

Experience is also important in reviewing plant design and participating in start-up and system check-out. While the O&M group does not begin actually running the plant until it is built, an experienced firm can increase the project's long-term operability and maintainability by being involved in the design and/or construction of a project.

For example, one plant superintendent saved a project $250,000 when he noticed that the exhaust duct gaskets planned for a project had a history of failing shortly after start-up. The superintendent suggested an alternate gasket and saved not only the replacement costs, but also the lost generation that the project would have incurred when the gaskets failed after the plant was operating.

An O&M company's experience is invaluable in critical decision making. When an emergency arises at a power plant, there is no time to do homework on how to handle the situation. Split-second decisions must be made, and there is no margin for error. The operator must be able to quickly evaluate a situation and act decisively.

For example, a Mission Operations and Maintenance employee recently noticed a pattern of alarms at one plant. Separately, the alarms did not signal a serious problem, but the operator notice that the pattern of alarms was similar to a set of alarms that preceded a catastrophic event at another plant. The operator tripped the unit, avoiding any damage to the plant.

So, in this case, experience was valuable not only because the operator did the right thing, but he was able to recognize the problem because a detailed analysis of the previous event was circulated to all plants with similar equipment. Thus, an experienced operating company benefits from proliferating its collective experience.

Lastly, but possibly most relevant, experience is important in planning. In the event of an equipment failure or other loss, valuable information can be gained from a post hoc analysis of the sequence of events leading up to the failure. Root cause analysis must be conducted on all significant operating incidents to determine an immediate fix as well as to prevent the failure from recurring.

At one plant last year, a series of redundant, "fail safe" systems failed in a way that had not been anticipated by the turbine manufacturer. The result was an equipment failure which destroyed three stages of the combustor section.

Within hours of the failure, the operators began to disassemble the unit and analyze the failure. By the next day, they used the information they had gained to change operating procedures and communicated this to other sites so the failure would not occur elsewhere. They accelerated the maintenance schedule and performed major repairs on the unit.

As a result of the analysis findings, we were able to work with the turbine manufacturer to redesign the turbine components and preclude a similar failure in the future.

### Standards and Practices

Just as crisis response is important, so is the discipline of an organization. Day-to-day, routine practices such as logging, passing the shift, use of standard O&M procedures, and safety are good indicators of an organization's commitment to quality service. In performing due diligence audits, we have seen some plants that operate under very lax standards which endanger the plant's long-term profitability.

For example, some plant operators "jumper," or disconnect, alarms such as temperature spread alarms. This can, in the short run, increase capacity factors. However, the equipment ultimately will suffer a failure.

One plant we looked at had jumpered its temperature spread alarms. This facility had some of the highest capacity factors we had ever seen. However, two years into operation, the turbine suffered a catastrophic failure, resulting in a major expense, *and* expensive down time that reduced its average capacity factor below industry standards.

**Personnel**

The only true asset an operation and maintenance services company has to offer is its personnel. The quality and productivity of personnel will be a function of the selection process and training regimen.

The most important component to obtaining quality operations is the hiring and training of quality operators and supervisors. The O&M organization should expend significant effort to use objective selection techniques, such as math, reading, and mechanical aptitude tests for operators.

Once hired, all new operators should be given training in the basics of electricity, plant chemistry, physics, electrical protection, and turbine operation. Electrical protection is an area of particular importance. This should be done both in the classroom and on the job. Some kind of check-off system for skills and proficiency training is a must.

Supervisory selection is also critical. We have done extensive research to develop a method for choosing supervisors. The characteristics and traits of a good supervisor were determined from indicative behavior of people with these traits. This methodology has been successfully implemented in an objective interviewing process which identifies people with these traits.

We have also developed skills programs to teach supervisors to perform specific tasks, such as conducting a performance appraisal, and competency programs which teach supervisors how to become key members of management.

Productivity is another indicator of the quality of an organization's personnel. It is our philosophy to leverage personnel time by broadly defining jobs and training operators to do maintenance work. We also get more out of our contractor relationships by competing for maintenance work with our vendors.

Case in point: the time required to perform a combustion inspection was cut in half by putting plant personnel on the job with vendors, rather than using only vendor personnel.

**Other Organizational Capabilities**

Finally, an evaluation of an O&M services provider is not complete without examining the company's other organizational capabilities. The firm's ability to review project costs, audit plant operations, review plant operation, safety, and maintenance logs, conduct due diligence audits, and conduct performance testing are indicators of the organization's overall depth and level of experience.

# OTHER FACTORS TO CONTROL RISK

When a lender considers an investment, the objective is to control risk. The single factor that can reduce the risk in the critical area of power plant operation and maintenance is the selection of an established, experienced O&M services company with a reputation for running plants for long-term profitability. O&M companies who operate and maintain their own projects are good candidates, because they tend to apply the standards they use in running their own plants to the operation of other plants.

Second, use incentive provisions in O&M contracts. If a company is willing to be measured against production standards and base some of their compensation of performance, their incentives are in line with the incentives of the project's owners. A capable O&M company will usually agree to a lower service fee with a potential to make more based on production performance.

Lastly, look for a company that is willing to take a long-term contract. Such companies are interested in the long-term success of the project to support its business base. This shows a commitment to operation and maintenance as a primary service, not just as a side business.

*Chapter 12*

# Cogeneration Facility Audits
# For Safety and Performance

*Chapter 12*

# Cogeneration Facility Audits For Safety and Performance

*Winter Calvert, P.E. and Kellie Byerly*

**M**any organizations that operate cogeneration facilities have developed an internal review and audit process for detecting and correcting potential problems. The purpose in this is two-fold. First, to detect any safety problems that might harm equipment or personnel. And second, to operate the equipment at the highest possible level of performance and availability.

The information in this article is not intended to qualify a facility for regulatory or safety compliance, but is presented as a typical example of the issues and programs involved for operating facilities.

For operators that are interested in improving both the safety and performance of their cogeneration plant, this introduction to facility audits will be a starting point. It is important that management at each facility explain how identified problems are not directed at individual employees, but are addressed as a team effort. This will help ensure that problems are corrected in an appropriate and timely manner.

There are five basic areas that cover most issues within a plant audit. These are:

1) Safety
2) Maintenance
3) Operations
4) Training
5) Administration

# SAFETY

Safety is paramount in any operating facility. It directly affects the employees' confidence in their work, and their abilities. It also affects the long-term profitability of a facility. Safety is an issue which must be understood and acted upon by each and every employee, and the importance of it emphasized by management on a daily basis. Most plants have safety meetings, at least monthly, to directly air concerns and discussion.

Each organization typically has a safety coordinator, whose primarily function is to implement procedures and follow up on corrective issues. This person should have a good understanding of OSHA as well as environmental terminology and procedures.

Safety training begins from day one with a new employee, and includes operational and recurrent training. There are various methods that can be used to inspect a facility for safety, but there is no single method that works well for every facility. It is important to tailor the process for an individual facility, taking into account plant age, type of equipment, experience level of personnel, etc.

Some sites have more difficult safety and operational requirements than others. The state of California has unique requirements that are generally over and above those required by other states. This also appears to be the direction in which the federal government is moving. For example, in California, Toxic Hotspots per AB 2588, are covered with an Emissions Inventory Plan (EIP) and an Emissions Inventory Report (EIR). Similar legislation is already being proposed by other states. Therefore, the audit should be tailored to local requirements (Table 12-1).

### Table 12-1. Basic safety inspection.

1. Safety Meetings
2. Safety Procedures
3. Hazardous Material Management (HMM)
4. Fire Protection
5. Compliance Reporting
6. Safety Equipment
7. Risk Management Prevention Program (RMPP)

# MAINTENANCE

There are three types of maintenance at any facility. It is important to have programs that address all three types, because each targets a different aspect of maintenance. These are:

1) Preventive Maintenance
2) Corrective Maintenance
3) Predictive Maintenance

**Preventive maintenance** includes such tasks as routine lubrication, borescope inspections, and instrument calibrations. This type of maintenance is intended to reduce and prevent the occurrence of equipment failure. It is usually scheduled and assigned to personnel on a regular basis, so that when unusual changes occur, they are captured before any damage is incurred.

Savings for this maintenance can be traced to the owner's bottom line, and the costs can be translated into operating profits. There have been several technical papers written which showed a direct relationship between preventive maintenance and unit availability[1,2] In one, an increase in plant availability the following year was demonstrated.

**Corrective Maintenance** are those tasks which follow a mechanical or electrical failure. This can include troubleshooting, engine changes, and repairs. Corrective maintenance is by far the most costly type, and is the reason Predictive and Preventive maintenance techniques are used.

**Predictive Maintenance** is a relatively new tool for plant managers. It includes a growing field of equipment diagnostic hardware that can be used to analyze and prevent failures before they occur. Like Preventive Maintenance, it is a tool which can have a direct effect on the bottom line. It is more analytical in nature, however, because it applies to those areas that are not accessible by traditional means.

For example, lube oil sampling and analysis, rotating machinery diagnostics, and instrument histograms are used in Predictive Maintenance.[3] Histograms are software-based programs that record and display an instrument reading over an extended period. They are particularly useful in tracking events prior to equipment failures.

All major equipment in a plant should have a maintenance activity log which describes a detail history of each failure, troubleshooting, and maintenance performed. All of the systems are audited for condition

(Table 12-2). Equipment that develops a history of problems and faults can be upgraded, using the maintenance activity log to assist with the financial justification.[4] Each type of maintenance in a plant can be audited for performance and improvements. These areas can usually be improved at a facility, and have a resultant savings.

**Table 12-2. Typical equipment condition audit.**

| | |
|---|---|
| 1) Gas Turbine | 9) MCC/Switch Gear |
| 2) Heat Recovery Steam Generator | 10) Acid Storage System |
| (HRSG) | 11) Caustic Storage System |
| 3) Steam Turbine | 12) Auxiliary Boiler |
| 4) Main Generator | 13) Refrigeration Unit |
| 5) Water Treatment System | 14) Diesel generator |
| 6) Compressed Air System | 15) Firepump |
| 7) Fuel Supply System | 16) Black Start System |
| 8) Fuel Handling System | |

# OPERATIONS

Operational reviews involve aspects of daily and routine plant activities. Some can be related to maintenance and others involve the activation and deactivation of certain plant functions. Are these activities scheduled in a logical and consistent way?

Operations also include daily record keeping functions which must accompany any facility, such as the Operating Log, visitor entry, scheduling of maintenance activities, monthly and annual plant availability, etc.[5] These procedures should be consistent and well-understood by personnel. Procedure manuals are a necessary part of operations. They are critical to safety and useful for consistent operations from day to day.

Plant procedures such as cold start-up and shutdown, and procedures for cold start and shutdown of the various systems are a necessary part of the procedure manuals. These include critical valve positions and order of events. Specific procedures should be included for each of the major systems, including the gas turbine, HRSG, boiler feed pumps, etc. Operations in icing conditions should be included in the standard operations manual.[6]

Utility outage logs and pass down orders should be available to the plant operator at all times. These are usually kept close to the operational log. A complete set of plant drawings and vendor manuals should be kept orderly and accessible.

# TRAINING

There are three types of training that are important to every facility:

1) Operational
2) Safety
3) Recurrent

A program should be developed for each facility that can be reviewed by management on a regular basis. **Operational training** involves a basic understanding of the function and maintenance of equipment, operational procedures (which is also safety related), and other procedures.

**Recurrent training** is a review and advancement of previously received training. Refresher courses are frequently overlooked at facilities because it is assumed that employees are knowledgeable about systems they have worked with for extended periods. However, everyone can benefit from refresher training. It is a good opportunity to, not only review basic material, but underscore their importance.

A list of subjects that might be included in a typical safety and regulatory training program is shown in Table 12-3.

**Table 12-3. Typical safety & regulatory training program.**

1) CPR
2) Confined Space Entry
3) Respiratory Protection
4) First Aid
5) Electrical Safety
6) Fire Prevention Program
7) Hazardous Materials
8) Emergency Response Plan
9) Risk Management Prevention Program (RMPP)
10) Hearing Conservation
11) Injury and Illness Prevention

12)   Process Safety Management
13)   Lockout/Tagout Procedures
14)   Blood-Borne Pathogens

# ADMINISTRATION

Reviews of the administration area include issues such as personnel files, purchasing records, cost controls, etc. Documentation of overtime, medical files, tracking of employee vacation time, shift scheduling in a timely manner, are all items that can be included in the review process. Safety and accident records should be tracked accurately.

Each facility can develop and implement some form of internal control that will fine-tune each of these areas. The results that can be expected are: improved plant availability, reduced accident rates, and improved profitability. Every facility is unique and has different considerations and requirements from other sites. Those operators that are best able to provide improvements in performance and availability, will best be able to accommodate the coming changes in the competitive field.

The authors gratefully acknowledge assistance from the following individuals: Don Wallin, Steve Huval, Jack Patton, and Robert Baten.

**References:**
1. Spector, R.B., "A Method of Evaluating Life Cycle Costs of Industrial Gas Turbines," *Journal of Engineering for Gas Turbines and Power*, Transactions of A.S.M.E.; New York, New York, October, 1989.
2. Simpson, Welton and Stoll, Harry, "The Influence of Maintenance Spending on Generating Unit Availability Performance," *Turbomachinery International*; Norwalk, Connecticut, July, 1989.
3. Calvert, Winter, "Gas Turbine Oil Systems: Operation and Maintenance," *Turbomachinery International*; Norwalk, Connecticut, 1995.
4. Calvert, Winter and Kleen, Randy, "Upgrade your Gas Turbine: Weighing the Costs and Benefits," *International Power Generation*; Redhill, Surrey, United Kingdom, 1995.
5. Peltier, Robert and Swanekamp, Robert, "Aeroderivative Engine Reliability and Availability," *Cogeneration and Competitive Power Journal*; Lilburn, Georgia, Winter 1995.
6. Calvert, Winter, "Prevent Damage to Gas Turbines from Ice Ingestion," *Power*, McGraw-Hill; New York, New York, October, 1994.

*Chapter 13*

# How to Evaluate the Economics
# Of an On-Line Cogeneration System

*Chapter 13*

# How to Evaluate the Economics of an On-Line Cogeneration System

*Anthony Pavone, P.E.*
*SRI International*

O nce a cogeneration system is up and running, a necessary engineering and business function is to assess how well it is performing, not only from a technical basis, but also from the perspective of a business decision. Is the project actually saving as much money as it was supposed to save? If there are deviations, and there almost always are, what are they?

This chapter describes some of the basic issues that need to be addressed once the plant is on-line, and the order in which they should be addressed. Also discussed: how the results of that assessment can be used to improve the economic and technical performance of the cogeneration system, so that it adapts to the current operating environment, rather than the environment that existed when the project's design was frozen.

When a cogeneration project starts up, a lot of attention is always focused on the operating performance of the hardware. While that aspect is always important, major deviations in financial performance of the project are usually not associated with machine performance, but, rather, deviations in the technical and business environment in which the project performs.

At the time the project was approved, a number of key assumptions and forecasts had to be made about the future. On start-up day, the future has arrived, and the first order of business is to compare the assumptions and forecasts with today's reality. Only after this is done should we get into the machine to evaluate operating heat rate, exhaust flange temperatures, fuel consumption, and boiler output.

# ASSESSING
# THE BUSINESS ENVIRONMENT

The key forecasts made prior to project approval were invariably the price of purchased electricity from the local utility, and the cost of fuel which was used to meet the thermal requirements of the process plant.

Are those forecasts still valid? If not, the project's operating performance will not meet forecast, no matter how well the actual cogen project was implemented.

Additionally, is the actual cost of purchased fuel for the cogeneration plant equivalent to what was forecasted? If not, trouble lies ahead. Although developers of large cogeneration investments lock in long-term fuel prices to protect their investment, many smaller cogen units continue to buy fuel on the spot market. I have been involved in a cogen project appraisal where natural gas purchases for the gas turbine were projected to the $1.80 per million Btu. During January's freeze in the Northeast, this customer paid nearly $4.00 for his gas, and his cogeneration project lost money.

Although the average cost of electricity displaced by the cogeneration project may not have changed, the two biggest components may be much different during plant operations. These are: 1) fuel component of electricity purchases, and 2) demand charge for electricity.

The second item is especially important if the cogeneration project, when running, experiences any trips. Some utilities change their demand charge monthly, others annually. If a particular cogeneration project operates in a high demand charge environment, in which a single trip sets the demand charge basis for the next year, it will be difficult for the cogeneration project to make money for the customer. In this situation, adding redundant instrumentation to the cogen unit to avoid spurious trips can pay for itself in minutes.

# EXTERNAL LOAD FACTORS

Many cogeneration operators mistakenly believe that their project is a production unit; it is not. Instead, it is a utility which creates value only when its output can be utilized. If its output cannot be utilized, the more you produce, the more money you lose. This is especially true when a cogen project is designed to make 50,000 pounds of steam per hour for a process consumer, but on startup day the consumer needs only 25,000 pounds per hour.

Venting excess steam production from a cogeneration project is a big money sink. The astute cogen plant operator will carefully assess what the demand for electricity and steam is before he actuates the cogen plant startup sequence. A key here is to make sure the cogen unit has enough turn-down or turn-up flexibility to follow demand. If not, defer plant startup until that flexibility gets added in.

On the electricity side, most cogen plants can back out utility purchase at, say, 4 cents/kWh, but push excess generated kilowatts back into the utility grid at an avoided cost of only 2 cents/kWh. Any deviation in plant need will have major implications on operating plant economic performance. In fact, the plant operator may find that optimum economic performance is not the same as optimum technical performance, and should choose carefully how he intends to run his new unit.

# SUNK COSTS

The operating performance of a cogeneration unit depends very much upon sunk costs for the investment, over which the operator has very little control. There is a popular myth that project cost overruns become history at start-up time. The impact of overruns, or to be more optimistic, underruns, will very much affect every day in the operating life of a cogeneration project.

After fuel costs, the single biggest component in the economic performance of a cogeneration capital is depreciation charges. Any major deviation in actual plant investment will affect depreciation, and overall project economics until the plant is written down to zero value.

That same total investment number is the denominator for calculating plant return on investment, and will dictate to a great extent whether the project becomes an operating winner or an operating loser. Depreciation charges on a large cogeneration project often represent 0.7 cents/kWh. For smaller systems, it will range from 1.0-1.5 cents/kWh.

More frequently, cost overruns create cash flow impacts on the operating performance of a cogen plant. These plants require a considerable amount of maintenance, which in turn is most often dependent upon the actual investment in the ground, rather than the forecast investment at the time the plant's design was appropriated.

Startup day is an excellent time to revise the expected annual maintenance budget to reflect what actually got built, and what it actually cost. Maintenance costs run about 0.4 cents/kWh, so that a project with a 50% overrun takes a 0.2 cents/kWh non-forecast penalty.

## OPERATING PERFORMANCE

Up to this point we have discussed those factors which affect the operating performance of a cogeneration plant before the unit is even started up. Now, let's turn to those parameters which do deal with a running plant. We will start with the hardware making kilowatts, and later turn to the equipment recovering thermal product.

## UNSCHEDULED OUTAGES

The single, most important thing a cogeneration plant operator can do to maximize the value of his investment is to keep the unit from turning itself off once it gets turned on. Trips not only wreck cogeneration economics, they wreck hardware and sometimes people.

What is the impact of trips outside the cogeneration battery limits? What happens inside the plant?

As mentioned earlier, an unscheduled outage can trigger a utility demand charge which lasts for a year or longer, wiping out this year's profits. Beside purchased power demand, another result is the purchase of backup power while the cogeneration plant is down. Somebody also has to supply the thermal load which previously came from the cogeneration plant.

When a cogeneration plant goes down, it sometimes takes the rest of the plant with it. The transition from cogen power to utility power creates a bump in the curve, and can knock out delicate instrumentation, as well as computers and other "power quality" consumers.

Unlike an on-purpose steam boiler which will continue to steam for a couple of minutes after the fuel control valve shuts off, the waste heat boiler behind a gas turbine will go dead in about 30 seconds. Maintaining the integrity of a plant utility steam system during a cogen trip is not a trivial matter.

The economic losses to the plant served by the cogeneration system will probably be a magnitude greater than the losses to the cogeneration system itself.

Better than half the trips of an industrial scale gas turbine cogeneration system are spurious instrument trips. This suggests very strongly the wisdom of redundant instruments.

From a hardware point of view, the worst time to be around a cogeneration system is when it is starting up or shutting down. Most operators know a great deal about the steady-state performance of machinery, but their transient performance is another matter. Not only is the wear and tear on equipment the greatest at this time, but the danger of a boiler flame-out, or turbine blade rub is greatest when temperature gradients are the largest during this same period.

# ROTATING MACHINERY
# STEADY STATE PERFORMANCE

The most important single parameter in the steady state performance of a piece of rotating machinery cogen equipment is the amount of kilowatts coming out divided by the amount of fuel going in. If the plant is on-target with this parameter, most everything else is under control. One occasional exception is being on-target in heat rate, but not being able to meet total expected output. Often this can be traced to an undersized fuel control valve, or obstruction of some type in air flow.

If the machine cannot meet heat rate, there are many possible problem areas. Inadequate airflow is a likely culprit, as is excess backflow pressure from the exhaust. An analysis of the exhaust gas composition and temperature can be used to diagnose incomplete combustion, fuel contamination problems, and faulty instrumentation.

Economically, an engine-based cogeneration system should consume 10,000-10,600 Btu of fuel (LHV) for each kilowatt produced. A gas turbine less than 5 MW should consume about 11,800 Btu/kWh, while a high efficiency industrial scale machine above 10 MW should consume 10,600 Btu/kWh.

Large scale, state-of-the-art industrial gas turbines, and aero derivative units, have heat rates below 10,000 Btu/kWh. Each supplier provides detailed performance data about his equipment. Actual deviations from this performance level can reduce economic performance of the cogeneration project by up to 0.3 cents/kWh.

It is very important to begin measuring all the performance parameters right when the system is started up for the first time, to be able to differentiate design and installation problems from those problems that are created by continued operation of the machine. Most machines, both engines and turbines, are run on a test stand before being factory shipped, and startup time is the best time to compare actual on-line performance with that recorded at the factory. On-line commercial performance should be stewarded against factory test-stand data for the particular machine, rather than general published data, or vendor guarantee data.

After initial steady state performance is attained, the machine should also be run through its complete operating range *while still new* to create the standards upon which all subsequent operating performance will be measured. When throttling down, the operations team needs to know where the surge point is, in order to avoid it.

Also, harmonic regions not identified by the factory need to be found while the machine is being attended. Many cogen systems are designed to run without human operators, requiring that potential problem areas be found during start-up.

# DEGRADATION OF
# STEADY STATE MACHINE PERFORMANCE

Over time, all machinery will degrade. Engines do down gradually. Gas turbines drop to about 80% of startup performance within a couple of days, and then go down slowly.

The best way to keep the rotating machinery from degrading prematurely is to keep the lube oil system clean, full, and cool. Monitoring lube oil temperature off the machine, sump levels, and occasionally analyzing oil samples can identify premature wear areas in bearings and seals. Large equipment is usually fully instrumented with accelerometers and other vibrating sensing facilities, and sometimes comes with a signature analysis to detect degradation.

Second only to the lube oil system is the combustion air system. Problems due to clogged air filters and dirty air compressor blades usually show up as poor heat rate, low total power output and high exhaust gas temperatures.

The economic impact of a poorly operating rotating machine making kilowatts can result in a penalty up to 0.3 cents/kWh.

# WASTE HEAT RECOVERY SYSTEMS

Most cogeneration systems recover waste heat as steam, some recover hot water, and a few exotic systems use exhaust gas directly to fire process furnaces. This discussion is limited to steam and water systems.

Gas turbine systems usually recover steam, because turbine exhaust temperatures range from 750-990°F. Engines usually recover hot water because their exhaust temperatures usually range between 400-550°F.

Many operating performance problems discovered in the heat recovery system are caused by problems in the rotating machinery. Low or high exhaust temperatures are usually due to fuel/air mixture problems, either due directly to fouling conditions, or due indirectly to poorly operating intake dampers at off-design load conditions.

Exhaust gas pressure problems can be due to the above, or to machine wear problems caused by excessive turbine blade clearance or piston ring blow-by. Unstable pressure cycling problems are often caused by air leaks at the transition piece.

Operating problems inherent to the waste heat recovery system can usually be diagnosed by performing an enthalpy balance around the boiler, with particular emphasis on the stack gas. High stack temperatures and high boiler backpressure are symptomatic of gas side finned tube fouling, and usually begin at the economizer.

If stack gas dampers exist, they can also be the cause of excessive pressure drop. Natural-gas-based cogen systems are the least likely to foul on the vapor side, but duel-fuel systems often foul when switching from gas to oil, especially if fin spacing was designed for gas only.

High stack gas temperatures accompanied by normal backpressure usually point to water-side fouling problems. Most water systems are designed for the proper treatment consistent with design steam pressure levels. However, some operators unintentionally create water-side fouling problems through inadequate blowdown from the steam drum, or failing to treat steam condensate before recycling it to the boiler.

A properly operating, unfired waste heat boiler system should generate one pound of steam for each 1200 Btu (HHV) of exhaust gas from the gas turbine or engine. In general, gas turbines are designed to generate 4-5 lbs of steam for each kilowatt produced. Firing the boiler with supplementary fuel will increase the ratio of steam production to electricity production. Recuperators and turbine steam injection can be used to increase the ratio of power output to steam output.

The overall benefits of cogeneration are in the "free" steam or hot water recovered from a power producing system. It should be obvious that this benefit deteriorates if that thermal energy is not economically recovered. Making steam or hot water is not enough. The recovered steam or hot water must be used somewhere in the plant efficiently to back out purchased fuel.

Too many cogeneration projects fail the economic performance test in operation because they fail to effectively use the waste heat. The most prevalent case is the inability of the process plant to use all the recovered thermal energy made by the cogen plant.

Whether this excess heat leaves in the form of high stack gas temperature, or as vented steam, the economic result is a poor cogen investment. I've seen at least one plant proposal to use cogen heat to substitute for waste bark in a paper mill. If implemented, the project would have lost a tremendous amount of money. Not only would cogen heat be used to replace free fuel, but the plant owner would have to pay extra money to remove all the waste wood products which were currently being consumed in the boiler.

# IMPROVING
# COGEN OPERATING PERFORMANCE

Although there is a fascination with cogen machinery, and a natural desire to optimize the performance of the equipment itself following initial startup of a system, a better place to start is the external environment served by the cogen plant.

Because the time lag between design of a cogen system and startup is anywhere from 2-5 years, much will have changed during the interim period. From the cogen operator's point of view, the two most important factors are the power load and steam load.

How much of the cogeneration system's design output can be economically used? Hopefully, all of it, and sometimes even more.

Often, the answer is *less*. The operator first needs to make sure that he operates his cogen plant to follow the required load. When changes in thermal load do not match changes in power load, tough decisions need to be made. Pushing excess kilowatts into the grid may or may not be economically wise. Venting excess steam is rarely economical.

Often, the optimum economic answer is to throttle the unit down to a point in which all the output can be effectively used, and possibly add some minor facilities to balance power vs. thermal output.

The next direction is avoiding unscheduled outages, especially nuisance outages caused by instrument trips. Redundant instruments help.

Next is an assessment of the most likely cause of real trips. This may mean adding facilities to prevent the level in the steam drum from swinging wildly as the cogen plant follows its load. Sometimes reinforcing the cogen support structure will damp out vibrations that might otherwise trip an engine or turbine.

Fuel system and lube system trips can be minimized by taking a hard look at the as-built drawings, and comparing them with the operating environment, rather than the design environment several years ago. In many cases, trip problems are caused at the interface between different suppliers' equipment, which is almost never reflected in design drawings.

Lastly, operating performance can be improved by looking at the routine operating parameters of the system. Meeting the manufacturer's guarantee is not good enough. All systems have some slack between what they can actually do, and what they are guaranteed to do.

The astute operator will determine those areas, and capture them in economic terms. This means studying the factory test results, and trying to achieve these in the commercial plant. When pushing a system to its output limits, it means finding out where the cliff is.

**Several years ago, beating the competition meant having a cogeneration system. Now that most everybody who can use them has them, beating the competition means making sure your cogeneration system is outperforming your competitor's cogeneration system.**

*Chapter 14*

# Environmental
# Impacts of Cogeneration

## Chapter 14

# Environmental Impacts of Cogeneration

*James R. Ross, P.E.*

T his chapter places in perspective the environmental impacts of cogeneration as a preferred alternative to other forms of heat and power generation with emphasis on waste heat and fuel as opposed to fossil fuel use. A methodology for deciding about cogeneration and the avoidance of environmental problems is offered.

## REFUSE BURNING

A major thrust currently underway in many countries is the burning of municipal waste as a means of producing energy in the form of steam, electricity or both. Energy production is a secondary purpose of these facilities commonly referred to as resource recovery plants. Some major success stories and some terror tales have developed from the increased use of this form of cogeneration.

Nashville, Tennessee, for example, heats and cools the entire downtown area by burning municipal waste in a thermal transfer plant. Although environmental problems, primarily odors, beset this plant in the early stages of operation, it is essentially a trouble-free operation at this time.

A smaller city, Gallatin, Tennessee, performs a resource recovery process for metal and glass removal, and burns the combustibles for process steam at adjacent manufacturing plants with the excess being used to generate electricity. This plant too had initial air pollution problems which have been resolved.

Some resource recovery plants have failed due to technical problems such as improper design, and others due to mismanagement or
unrealistic expectations. Others have failed due to public outcry, as no-
body wants a garbage-burning plant next door to one's home. Once the
perception of such facilities becomes negative, technical excellence and
environmental acceptability become inconsequential.

## Landfill vs Waste Burning

Landfills have become popular with environmentalists because
there is no air pollution involved. The battle cry of environmental groups
in the 70's was to stop burning. Any air pollution was intolerable, and
the only alternative offered was to put all waste into landfills. We are
only now beginning to see the folly of this bad alternative.

First, landfills are becoming filled to capacity in many locations.
Sites for landfills are becoming increasingly difficult to find as land with
suitable characteristics does not exist in many locales. Florida, for example, has few hollows, and is forced to build mounds of garbage on flat
sandy soil where the water table is about 4 feet below the bottom of the
landfill. Most of the waste in these landfills could be burned with little
impact on air quality.

Other landfills are located within a mile of fresh-water lakes from
which drinking water is pumped. The impact on water quality in the
next century may be devastating due to the unknown long-term effects
of leaching of plastics and other chemicals which may be greater in the
long run than a temporary air quality problem.

Technology has kept pace for the solution of air quality problems,
even though the solutions are too expensive to implement in many cases.
Advocates of landfills contend that technology will keep up, and that by
the time this gets to be a real problem, there will be a technological
solution. This same argument when used in defense of nuclear power
plants is loudly rejected by the anti-nuclear activists, many of whom are
also landfill advocates.

## How to Dispose of Waste

Waste is something that costs money. For example, packaging materials which end up in the trash have been paid for as part of raw
material cost. The beginning of the waste stream is typically far away
from the landfill or trash dock, and is really the result of a decision
process which begins in product design, purchasing, engineering and

even on the top manager's desk of manufacturing firms which are both customers and sellers of packaged goods.

**It should never be assumed that any waste item cannot be eliminated, utilized more effectively, or recovered.**

If the organization recognizes this concept, much of the waste can be eliminated at the source, and recovery efforts will be unnecessary. If the organization has a consistent value system and strategy aimed at waste reduction, much can be accomplished.

Paper and wood which represent almost half of most municipal and industrial waste streams can be burned with essentially no negative environmental impacts. Wood and paper smoke contains no sulphur, dioxins or $NO_x$, and while visible are essentially harmless.

Some plastics may be burned with little impact while others produce noxious gases which must not be emitted into the atmosphere. The best use of plastics is to sort and recycle them for conversion into specific recycled plastic or into a usable generic plastic rather than risking their decomposition in a landfill or burning them with adverse effects on air quality.

Metals should be segregated from the waste stream and not subjected to landfilling. Some landfills actually bury old washing machines, refrigerators, conveyors or other heavy metal objects which could be sold to a metal recycler. Generally, the landfill operator can profit from eliminating metals from the landfill and selling them to a recycler. The technology for segregating aluminum from steel is now developed, and if done, burning of trash can be made more efficient and profitable.

Communities which use curbside segregation of trash (metals, glass and plastic) not only get the value of the recycled materials, but also reduce the volume of materials which are passed through the incineration process, eliminate these materials from landfills and avoid negative environmental impacts of mixed burning or landfilling of unsegregated trash. Some communities which do not use curbside trash segregation use prison or welfare recipient labor for trash segregation to reduce the cost of the operation.

If waste burning includes plastics, tires or other substances which produce toxic fumes which may damage the ozone layer or create health problems for people, recycling is a preferred alternative. In any event, adequate safeguards must be taken to avoid damage to air quality.

While air quality is only slightly involved, further segregation of municipal waste should remove food waste, grass clippings and similar organic material which does not burn well, but which can be composted successfully, and the compost used for organic fertilizer.

## Hazardous Waste Burning

In some cases, volatile hazardous materials can be burned in an incinerator on-site and the heat so generated can be reclaimed, thereby qualifying the facility as a cogeneration installation. Hazardous volatile wastes are often blended in with other fuels, and can be burned with minimal environmental impact.

There is a whole continuum of options relating to the disposal of hazardous waste. Burning of hazardous waste is one of the less desirable options on the table as identified by Turner et al in Reference 1.

If a hazardous waste problem exists, the options of eliminating it at the source or recycling on-site are much more desirable both from an economic and environmental perspective. On balance, however, it is often less costly to burn hazardous volatile wastes on-site than it is to pay a disposal service to haul it off and dispose of it, depending on volume and type of waste. The technology for hazardous waste burning is rapidly catching up to the demands of industry.

## Electric Generation Using Fossil Fuels

Commercial electric utilities chiefly use coal, oil or natural gas to fuel electric generation. Anyone who listens to television news or reads a newspaper knows that burning of coal is a problem due to acid rain which is produced by sulphur dioxide and nitrous oxides mixing with rain water. Coal became a popular substitute for oil to fuel electric generation in the mid-70's due to oil shortages perceived in that era, even though many electric utilities had only a few years before been forced by the E.P.A. to convert to oil to comply with air pollution regulations.

This capital drain caused by switching fuels in quick succession has precluded implementation of many more attractive cogeneration alternatives which could have saved energy and money. Electric utilities could augment coal with waste fuels, and reduce the percentage of sulphur dioxide in the emissions to more acceptable levels.

A very viable solution is to mix wood waste or sawdust with coal. Another solution would be for the utilities to become contract waste-burning facilities.

## Nuclear Power Generation

The ultimate environmental impacts of nuclear power are yet to be determined. Nuclear plants are built to the state of the art and to the limits of knowledge which exist at the time of their construction, but this is continually changing. At the moment of generation, there is essentially no environmental impact from nuclear power generation; however, the impact on the environment may occur many years in the future if a solution to the nuclear waste problem is not found. The issue of nuclear waste is continually with us, and ultimate disposal concepts are still years away.

In building a nuclear plant, most companies must spend far more money than planned to comply with ever more stringent regulations and enforcement procedures. Little thought has therefore been given to opportunities for cogeneration at nuclear plants. Hopefully, as this industry matures, further effort to take advantage of opportunities for cogeneration in nuclear power production will yield positive results.

## Wood Waste Cogeneration

One of the bright spots in the cogeneration vs. environment picture is the wood products industry. Wood contains no sulphur or other injurious factions. It even smells good when burned, and the smoke can be burned without trace if temperatures are sufficiently high.

A very successful cogeneration installation was made in 1980-81 by a small wood products company. The manager was looking for ways to utilize more of the wood waste, much of which was being sent to a landfill or sold at a loss for paper mill fuel. He began investigating cogeneration, and since a boiler was already installed, the step to cogeneration was relatively easy.

Regulatory hurdles and contract negotiation with the utility had to be surmounted. As this was a new concept to the utility, a cogeneration policy had to be developed.

Using funds obtained from a DOE Appropriate Technology Grant, the client purchased a used steam engine and generator set, and proceeded to install the unit. New interface equipment was also installed. While some down time to get parts for the steam engine has been experienced, the client has received about $50,000 per year in revenues for electricity sold to the utility. This is just one of many cases where cogeneration from wood waste has been made profitable without damage to the environment.

A larger wood products plant installed large steam boilers, turbines, electric generators and pollution control equipment, and effectively isolated itself from the outside utility with the result that wood waste was being disposed of productively and electric costs were reduced significantly. A major chemical company installed a wood-fueled power station at its major facility, and obtains waste wood from a 100-mile radius of the chemical plant. Adequate pollution controls are incorporated into the power station.

Utilities which burn coal as fuel for electric production have experimented with blending wood shavings with pulverized coal with the result that the surplus content of emissions was reduced. In some cases the blending of wood with coal brought the total emissions within acceptable tolerances not being achieved with coal alone.

Some notable failures occurred in the late 70's in attempts to utilize wood waste through pyrolysis; however, these failures increased the body of knowledge, and will ultimately lead to future successes.

### Other Cogeneration Options

As research continues, other types of cogeneration will be discovered; however, there will be environmental objections to each, both valid and fictional. Engineers must continue to search for new solutions to energy and environmental problems with the objective of extending the life of this planet and stretching its resources.

### Bad Alternatives to Cogeneration

We have discussed the negative aspects of burning coal which produces acid rain, and the pollutants present in diesel exhaust emissions.

We have also mentioned the long-term negatives of landfills as a means of disposal of solid waste, but have also discussed the options of cogeneration, recycling and resource recovery which can make solid waste disposal environmentally acceptable.

Since 1981, oil has been plentiful, though sources of supply have been in jeopardy from a military standpoint. Oil prices have moderated, and the higher costs have been incorporated into corporate and individual budgets. The goals of energy independence so loudly espoused in the late 70's have been dropped due to this fortunate turn of events; however, the world's oil supply is limited (subject to political manipulation) and undependable. Cogeneration can reduce dependence on oil, and can do so with less impact on the environment than continued oil consumption.

Cogeneration is a viable option to the above bad alternatives, and it behooves industry and government to act decisively to promote cogeneration as an environmentally preferable alternative to wasteful or polluting practices.

### Considerations about Cogeneration

Organizations considering cogeneration should go through a rigorous decision process which would address corporate strategy, capital availability, costs vs. benefits, dependability of waste fuels or energy forms, environmental impact, public perception of the project, commitment by management and availability of management people to keep the system going long term. (See Ref. 2, 3.)

The following additional concerns should impact the decision about cogeneration:

A.  Determine first if all or part of the waste to be recovered can be avoided. If the process is designed properly, and the proper decisions are made about waste avoidance, perhaps the entire cogeneration project can be avoided. You don't have to cogenerate if you don't have the waste in the first place.

B.  Isolate any part of the waste stream, either material or gases which can be recycled in other ways to reduce the waste stream. If the waste stream can be reduced, the size of the cogeneration facility can also be reduced to handle what's left.

C.  Get the maximum benefit from the cogeneration and recovery effort by finding every applicable waste, and adding more input to the system as opportunities become available.

D.  Avoid damage to the environment by following steps A and B, and by installing environmental safeguards at every level of the recovery process.

## CONCLUSION

As with any innovative concept, cogeneration has some problems which must be resolved, including those associated with environment.

Too often critics of any new idea will see only the negatives, and will ignore the positives which far outweigh the problems. Where environment is concerned people, often with hidden agendas, may disrupt worthwhile projects because they ignore the positive benefits.

Despite some negative environmental impacts, cogeneration is an attractive alternative to other forms of heat and power generation. By using the same energy more than once, cogenerators actually reduce environmental impact as well as save energy, and the environment and the energy supply are both positively impacted. The net impact of cogeneration on the environment is significantly less than other forms of energy consumption and production.

Organizations wishing to cogenerate should consider the environment, and take all necessary safeguards. In some cases, an objective of cogeneration may be to improve the environment as well as save energy or reduce energy costs. Future cogenerators should accept environmental problems as opportunities, and should proceed to obtain the significant benefits which cogeneration can yield.

### References
1. Turner, Wayne C., Webb, Richard E. and Shirley, James M., "Management of Hazardous Materials, Chemicals, and Wastes. Case Studies Involving Large Savings," Institute of Industrial Engineers 1986 Fall Industrial Engineering Conference Proceedings.
2. Ross, James R., "Cogeneration—A Concept for Today." American Institute of Industrial Engineers Spring Annual Conference and World Productivity Congress, 1981.
3. Ross, James R., "Strategic Planning for Solid and Hazardous Waste Management and Resource Recovery," Institute of Industrial Engineers Integrated Systems Conference, 1987.

Chapter 15

# Environmental Permitting for New Non-Utility Electric Capacity

## Chapter 15

# Environmental Permitting For New Non-Utility Electric Capacity

*Robert J. Golden, Jr.*
*Envirosphere, Inc. Division of Ebasco, Inc.*

<br>

**M**ore and more electric utilities are relying on competitive bidding systems to supply additional electric capacity. Under the bidding system approach, the respective electric utility system sets out minimum economic, reliability, risk, operational and environmental criteria and procedures for procurement of electric capacity from cogenerators, small power producers, independent power producers and waste-to-energy facilities.

In New Jersey, for instance, bids submitted as part of a utility bid solicitation held pursuant to a Stipulation of Agreement for Cogeneration and Small Power Producers, assigned a maximum weight of 55% to economic factors, with minimum weights given to:

1) Project status and viability factors and

2) Non-economic factors being 25% and 20%, respectively.

Alternatively, in New York, the scoring factors for a Niagara Mohawk's recent bid solicitation (a maximum score of 1,310 for supply projects) include: price (maximum 850); economic risk (75); success (64); longevity (21); operational factor (80); and environmental (220).

In addition to bid solicitation criteria such as those detailed above, there are several other critical provisions of the bid solicitation that need to be thoroughly analyzed so as to prepare a more responsive bid and/ or to put the bid submittal in its proper perspective. These provisions revolve around the specified date for which capacity must be available to the utility; the provision for liquidated damages should the project fail to maintain the contracted amount of dependable capacity; and the requirement for the non-utility power producer to post earnest money (Virginia Power's solicitation in late 1989 required $30/kW), which is forfeited if the project is delayed or not completed.

Taken together, it is clear that non-utility power developers must adopt a comprehensive approach towards the development of non-utility generating capacity to avoid the significant penalty provisions included in a successful bid. Potential developers/turnkey contractors in every state must recognize the importance of properly identifying environmental permitting and scheduling requirements as well as construction and permitting time line requirements as early as possible. The risks for a developer who will be contractually committing to dates certain as to when various environmental and construction-related permits will be obtained as well as when power will be reliably generated and transmitted to the grid are enormous.

# DISCUSSION OF BID SOLICITATIONS
# AND DEADLINE FOR POWER AVAILABILITY

Tables 15-1 and 15-2 illustrate typical time frames for a bid solicitation from bid issuance to the deadline for receipt of electrical power to be generated by the alternative energy facility. By examining the time lines illustrated in the tables, one can see the critical importance of environmental permitting in the overall development of alternative energy projects. Several key permitting aspects of a typical project and methods of obtaining these permits in a timely manner to the project's overall development schedule are presented on the following pages.

# SELECTED PERMIT APPLICATIONS

There is no "standardized" permit check list for an alternative energy project. The local characteristics of the site selected, the regional

## Table 15-1. Consolidated Edison, New York; 200 MW Solicitation - Available by 5/1/94.

| | 1990 | | | | | | | | | | | 1991 | | | | | | | | | | | | '92 | '93 | '94 |
|---|---|---|---|---|---|---|---|---|---|---|---|---|---|---|---|---|---|---|---|---|---|---|---|---|---|---|
| | J | F | M | A | M | J | J | A | S | O | N | D | J | F | M | A | M | J | J | A | S | O | N | D | 5/1 | 5/1 | 5/1 |
| RFP Issued | | ● | | | | | | | | | | | | | | | | | | | | | | | | | |
| Notice of Intent to File Bid with Con Ed | | | ● | | | | | | | | | | | | | | | | | | | | | | | | |
| Pre-bid Conference | | | | ● | | | | | | | | | | | | | | | | | | | | | | | |
| Submittal date for Proposals | | | | | | | | | ● | | | | | | | | | | | | | | | | | | |
| Con Ed Selection of Preliminary Contract Award | | | | | | | | | | | | ● | | | | | | | | | | | | | | | |
| Negotiations Initiated | | | | | | | | | | | | | ● | | | | | | | | | | | | | | |
| Contract Negotiation Conclusion | | | | | | | | | | | | | | | | ● | | | | | | | | | | | |

24-month Construction Schedule, Testing and Commencement of Commercial Operations ←→

**Table 15-2. Jersey Central Power and Light Company, New Jersey 270 MW Solicitation\* - Available by 6/1/94.**

| | 1989 | | | | | | | 1990 | | | | | '90 | '91 | '92 | '93 | '94 |
|---|---|---|---|---|---|---|---|---|---|---|---|---|---|---|---|---|---|
| | M | J | J | A | S | O | N | D | J | F | M | A | M | 5/1 | 5/1 | 5/1 | 5/1 | 5/1 |

RFP Approval By BPU

RFP Issued

Bids due at JCP&L

Preliminary Award Group by JCP&L

Final JCP&L Selection

Contract Negotiation Period

30 day extention

Assume by 6/1/90

BPU Contract Approval

Preparation and Submittal to Securing all Enviromental Permits (by 6/1/92)

Construction, Testing and Commencement of Commercial Operations

\*Two of the projects selected by JCP&L were:
1) A 100 MW coal-fired plant sponsored by Fluor Daniel-Paulsboro Energy and
2)A 100 MW coal-fired plant proposed by Mission Energy-Vista Energy

aspects of the project, the details specified in the agreement with the thermal host and with the conceptual engineering design proposed, and the terms of the approved power sales agreement with the utility dictate differences in the magnitude and complexity of permits necessary to develop a project.

## A. Air Permitting

The key environmental permit in the permitting process in almost every case is the PSD/state air permit preparation/review. This permitting task represents the longest lead time and the one, more often than not, that is held hostage to ever-changing state agency air emission policies.

The driving force behind this situation is the consistent pressure being applied by state regulatory agencies to lower pollutant (principally $NO_x$) emission levels and the progress being made by equipment vendors in lowering these levels. While this situation leads to uncertainty with respect to acceptable emission rates, the chronic understaffing of air quality offices can lead to seemingly endless review cycles.

Should the project generate requests by local citizenry, environmental groups, etc., for more information or even outright opposition, the already lengthy review cycles become even more extended as gunshy agency officials seek to have all the t's crossed and i's dotted before issuing a permit decision.

One recourse available to shorten review time frames is to design the facility in such a way as to reduce ground level air quality impacts to insignificant levels. Stack height, proposed turbines, and/or control equipment, and sulfur content of fuels are variables which the developer must focus attention on in order to shorten review time frames. The time and cost involved to obtain a variance (if necessary) for stack height at the local level, for instance, must be weighed against the incorporation of control equipment such as selective catalytic reduction (SCR) or the use of low sulfur fuels.

Utilization of SCR technology in New Jersey also introduces the need to both comply with the Toxic Catastrophe Protection Act as well as respond to local municipalities' safety concerns with respect to the transformation, handling, storage and use of ammonia.

## B. Water Supply

The availability of an adequate water supply is critical to the operation of a cogeneration facility. The magnitude of water supplies needed for a project is, of course, dependent on the size of the project, the use of the water (e.g., water injection, cooling) and the source of the water.

In many instances, the availability of an adequate water supply has become as critical an issue to the successful development of an alternative energy facility as that of satisfying air quality permitting concerns. Concerned citizens and environmental groups, as well as local, county and state officials, are becoming ever more cautious with respect to water allocation permit decisions, especially when these decisions are being made under potential drought emergency declarations and water use restrictions, which in many regions have become commonplace during the summertime.

Decision makers are also becoming increasingly concerned with the potential for contamination of water supplies as the groundwater plumes from contaminated site areas are potentially altered from the requisite water drawdowns.

Depending on the proposed cogeneration facility's location, there are a variety of mechanisms through which an adequate water supply can be obtained. For illustrative purposes, some parameters for a proposed cogeneration project that is to be developed on leased land have been set out. The proposed facility, which would be developed by a third party on leased land, could run on the host site's well water supply entirely, city water, or both.

Further, the plant developer has the host site's conceptual approval for the use of its well water to be withdrawn under the host site's existing water allocation permit and there has also been an agreement signed affixing a monetary value to each gallon of well water to be used. The agreement, however, does not specify minimum-gallon allotments.

With these parameters set, if the facility should operate 100% on well water (as opposed to 100% on city water), the developer would realize a savings of greater than $50,000/year. In contrast to the cogeneration facility's water needs, the host site's corporate planning group has on the drawing boards a plan to construct a major new production facility at the same site location which will require considerable water volumes. The obvious source for the host company to obtain such water supplies is the well water which otherwise would be provided to the cogeneration facility.

Less desirable alternatives include amending its existing water allocation permit or drilling new wells. In New Jersey, for example, one can now expect to have to conduct geohydrologic studies and a 72-hour pump test for any water allocation requests greater than 100,000 gal/day.

The cogeneration plant owner has the option of doing nothing or utilizing the host site's well water until it is no longer available and then incurring the higher costs associated with utilizing a city water supply. Alternatively, the cogeneration plant owner could commence negotiations with the water utility for a more cost-effective arrangement or it could seek permission from the host site to drill new water wells on the leased property or on the adjacent host site's property.

As you can see, each alternative has an associated cost and time requirement. The alternative energy facility developer needs to carefully weigh the short-term versus long-term benefits and costs associated with each option.

### C. Sewer Extensions

The third example presented here for discussion revolves around a "minor" permit application—a sewer extension permit. This permit, in most cases, should not take more than 3-4 months to be issued.

For all intents and purposes, this permit application is relatively straightforward. However, there is a pitfall to potential cogeneration developers/turnkey constructors in obtaining this permit in many municipalities. These municipalities are the ones which are operating under a sewer moratorium, either self-imposed or mandated, which prohibits the physical connection of new sewer hook-ups to the existing municipal system. Until such time as corrective actions can be taken or completed, the bans will stay in place. For some municipalities, these corrective actions will not be completed for several years.

There are alternatives to circumvent new hook-up moratoriums but they can be extremely costly to implement or involve considerable time to satisfy appropriate officials. Thus, the lesson to be learned is that the developer who is submitting a detailed permitting schedule to a utility in the bidding process must do some homework to avoid the pitfall of setting out an unrealistic and/or unachievable permit schedule.

### D. Construction Permits

In many power sales agreements approved by State Public Service Commissions (PSC), there are critical milestone dates specified that the developer must meet. One particular milestone is the date by which significant construction work (e.g., more than site clearing and grading) has commenced. The implication of meeting such a critical milestone is that a turnkey design/construction contractor needs to be retained months beforehand.

More importantly, it can be seen that the time constraints imposed by competitive bidding systems, as detailed in Tables 1 and 2, are more likely to force the cogeneration developer to retain an engineering design/construction contractor earlier than preferred so as to allow construction deadline dates to be met. At the same time, however, the actual construction bid accepted by the developer more than likely does not reflect detailed site information contained in previously submitted permit applications or the provisions of already approved permit applications.

## CONCLUSION

Alternative energy facility owners and developers know that environmental permitting is an expensive and time consuming proposition. They are also now recognizing the importance of ensuring such environmental compliance, especially as they become aware of the significant role which environmental factors/feasibility play in the evaluation of the competitive bid as well as the role such factors play in allowing for timely completion and dependable operation of the facility.

## RECOMMENDATION

The potential ramifications to the developer/turnkey contractor's return on investment (i.e., the bottom line) of not properly recognizing the importance of environmental permitting/compliance can be quit significant. In an attempt to avoid such a scenario, alternative energy facility developers or turnkey engineering design/construction contractors who assume all permitting responsibility as part of their contractual agreement with the developer should consider the following:

1.   As early as possible, a licensing feasibility/fatal flaw analysis should be performed. This study would serve to identify potential

pitfalls in the licensing process; define an approximate licensing schedule; and suggest design changes which could expedite the licensing process. For the developer, key environmental parameters/critical paths can be identified and reflected in the capacity bid to be submitted to an electric utility.

2. Each component of the project and responsibility for permitting each aspect of the project must clearly be delineated. For example, if interconnection facilities are needed to allow the power generated to be transmitted to the electric power grid, it must be clearly specified as to whether it will be the electric utility's responsibility to permit such infrastructure or that of the cogeneration developers.

3. Never underestimate the extent and complexity of what is required and necessary to receive local planning board and/or zoning board approval. The local permitting aspect of a project is a "wild card." Do your homework to ferret out what the real local concerns are (e.g., air, noise, water, aesthetics, safety concerns, etc.) and then set out a workable action plan to address such concerns.

4. Establish an effective level of communication between all elements of the project "team." Often, there are two to five firms associated with preparing materials for various aspects of the project. Timeliness and responsiveness in preparing individual permit applications are predicated on receipt of accurate data and diagrams from each team member. Without effective communication between team members as to what is needed or correspondence between members indicating that certain aspects of the project have changed, there is the likelihood that valuable permitting time will be lost, not to mention the increase costs involved.

5. The developer must become more cognizant of the engineering/ design information needed early in the process to support the preparation of various permit applications even though it is a foregone conclusion that the design and configurations of the proposed plant will change over time. These changes can be attributed to a number of reasons—cost effectiveness, vendor guarantees or lack thereof, regulatory agency requirements, etc. It is important to keep

in mind, however, that these changes can also result in the need to modify previously filed permit applications or require the filing of additional permits.

In the overall scheme of things, environmental permitting is just one component of a long, involved process. However, proper attention to this particular aspect of project development will facilitate the securing of all necessary permits and approvals at all levels in a timely manner. More importantly, it will also have a significant positive impact on a project's bottom line.

*Chapter 16*

# Peak Shave/Base Load Cogeneration System: A Way to Improve Cogeneration Economics

*Chapter 16*

# Peak Shave/Base Load Cogeneration System: A Way to Improve Cogeneration Economics

*Luco R. DiNanno*
*Tecogen Inc.*

T he Peak Shave/Base Load Cogeneration system is an outgrowth of Tecogen Inc.'s packaged cogeneration systems technology. In the smaller sizes (i.e., 30, 60, and 75 kW) these systems utilize an automotive-derivative natural gas engine rather than an industrial gas engine. In addition, induction generators are used to minimize switchgear costs and the units are equipped with microprocessor controls and remote monitoring capability to improve maintenance economics.

At the 60-kW rating, the 454-CID engine has been shown to operate in excess of 20,000 hours with only routine maintenance. Its cost at that power rating, in dollars per kilowatt, is roughly one quarter that of an industrial engine.

With this very positive experience, it seemed likely that the 454-CID engine could operate at significantly higher power levels for shorter periods of time while still maintaining good operating economics. This conclusion led to the concept of a combined cogeneration and electrical peak shaving system, which would further enhance the overall economics of the system relative to a conventional cogeneration system and open up a broader market.

# TWO-SPEED GENERATOR MOST PRACTICAL

While there are a number of approaches that could be taken in exploiting the concept of a dual-power level system, a two-speed generator was adopted for this concept as being the most practical for a near-term, two-power level product. A simple four-pole induction machine could be built with external connections which would allow either two-pole or four-pole (3600 rpm or 1800 rpm, respectively) operation. The engine(s) could be directly connected to the shaft of the generator, thereby eliminating the necessity for a clutch and gearbox.

The peak shave/base loaded cogeneration system selected for development consists of two 454-CID engines coupled to a single two-speed induction generator equipped with a double ended shaft. Since the engine is available in either rotation, this arrangement poses no problem in modifying the engine. At the two operating conditions, 1800 and 3600 rpm, the goal is to have each engine operating at 80 and 160 kW's, respectively.

There are two reasons for selecting an induction generator for this application. The first is low first cost. This, of course, determines that the unit cannot operate as a standby or emergency generator. The second reason is that a brief inquiry as to the feasibility of a two-speed synchronous generator showed that no such product is available and that significant development effort would be required to produce such a machine.

At the present time, a laboratory version with a system rating of 160 kW at 1800 rpm and 320 kW at 3600 rpm has been built for development and testing at Tecogen, and a second prototype field experiment unit has been built for Baltimore Gas and Electric to be tested at the Marriott Hunt Valley Inn in the Baltimore area.

An overall system schematic for the naturally aspirated version is shown in Figure 16-1. Here the double lines indicate exhaust gas plumbing, the heavier single line shows lubricating oil system plumbing, and the lighter single line shows coolant system plumbing. The two engines are shown bolted to either end of the generator through flywheel adaptor housings.

The engines and generator form a rigid beam which can also be seen, in Figure 16-2, a photograph showing the side view of the field experiment unit. An end view of the unit is shown in Figure 16-3. The engines are equipped with water-cooled exhaust manifolds to improve heat collection efficiency and minimize thermal radiation to the sur-

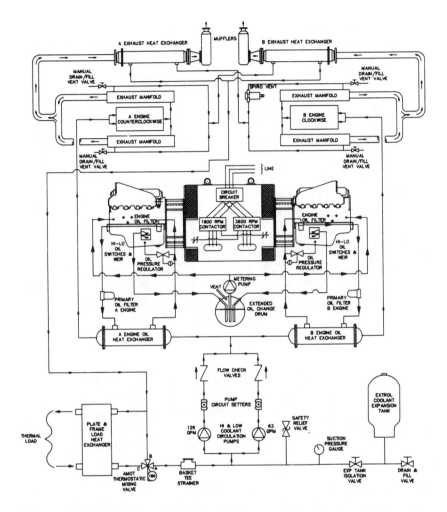

**Figure 16-1. Peak Shave/Base Load Cogeneration System Schematic**

roundings. All of the heat collected from the system is rejected to the load heat exchanger which is part of the packaged system. The user's thermal load sees only the colder side of this heat exchanger.

As Figure 16-1 shows, heat rejected by the engine lube oil system, jacket, exhaust manifolds, and exhaust heat exchanger is absorbed by the coolant and delivered to the load heat exchanger and a thermostatic mixing valve which controls the temperature of the coolant returning to the system.

Figure 16-2. Side View, Peak Shave/Base Load Cogeneration System

Figure 16-3. End View, Peak Shave/Base Load Cogeneration System

# THERMAL LOAD VARIATION

Most of the thermal load side of this system is like any other cogeneration system but is different in one significant respect. Because the engines operate at two very different loads (and speeds) the heat rejection of the engines at the two conditions is quite different. A motor-driven pump circulates coolant through the entire system. It is desirable to keep the coolant temperature into the engines and the temperature rise across them relatively constant for reasons of engine durability. To accomplish this, two different flow rates are provided by two main circulating pumps.

The lubricating oil system is unique in several ways. The oil change interval is extended by the 50-gallon reservoir which recirculates oil with the main engine sumps at the rate of about one gallon per hour. This system, together with the engine filters and a separate, very large main oil filter for each engine, should allow at least 2,000 hours and up to 4,000 hours between oil changes, depending on the amount of time the unit operates in the peak shaving mode.

The module containing all of the plumbing and heat exchangers can be seen under the engine and generator assembly (Figures 16-2 and 16-3). The heat exchangers provide longitudinal stability to the module, which consists of 4 ribs spaced one under each engine and two under each end of the generator. This construction technique eliminates the use of heavy subframe channel or I-beams which are basically unnecessary since the engine-generator assembly is already a very stiff beam.

The engine-generator assembly is mounted to the tops of the subassembly ribs through vibration isolators. Connection to the plumbing from the engine is through easily disassembled fittings. The electrical control box (shown in Figures 16-4 and 16-5) is mounted to the assembly and contains the control system and all of the switchgear and other electrical equipment.

The control system of the peak shave/baseload system uses a new single-board controller hardware platform based on a Motorola 68000 series chip which is being introduced across the Tecogen product line. The control system shown mounted to the electrical control box door in Figure 16-5 incorporates a complete operating system and is programmed to operate the peak shave/baseload system automatically, including the engine speed switching function.

Figure 16-5. View of Control System and Switchgear in Electrical Control Box, Peak Shave/Base Load Cogeneration System

Figure 16-4. Electrical Control Box, Peak Shave/Base Load Cogeneration System

The controls also monitor the operation of the system and store operating data for retrieval and analysis. The control system permits either on-site or remote operation of this machine and data retrieval and provides the means for remote dispatching of peak shaving capability.

# A DIFFERENT CONTROL LOGIC

There are basic differences between the peak shave system control logic and the typical logic of other Tecogen cogeneration systems which are programmed into our controller. These are:

1)   After the start mode is completed and the unit is "running," a check must be performed to ensure that both engines are operating. A comparison of actual idle speed to an expected idle speed with both engines operating is made by the controller; lower than expected idle speed results in shutdown and a starting error message is displayed.

2)   A means of equalizing the power output of the two engines is required. The approach adopted is to use a stepper motor actuator at each engine and to operate them in tandem during transient conditions. At full load at either speed, the controller reverts to independent control of the two throttle stepper motors with one becoming the "lead" engine attempting to hold full power and the second becoming the "follower" engine and attempting to reduce the difference in manifold pressure between itself and the lead engine to zero.

3)   A means of changing the operating mode while the unit is running (as opposed to shutting down and restarting in the new mode) is required in this system. The possibilities are
a)   Shutdown—No heat or peak shaving required
b)   Go to peak shave mode
c)   Go to cogeneration mode

This situation is easily adapted into our controller. The new logic has been developed and demonstrated.

The entire unit may be lifted with a forklift truck from underneath or by an overhead crane through the generator. Table 16-1 gives the preliminary specifications for the naturally aspirated version of the Peak Shave/Base Load Cogeneration system.

**Table 16-1. Tecogen Peak Shave/Base Load Cogeneration (PS/BLC) System with Naturally Aspirated Engines.**

<u>Preliminary Specifications</u>

Output
- Cogeneration Mode at Engine Speed of 1800 rpm
  -Electrical -160 kW, 460 V, 3 Ph, 60 Hz, Power Factor 0.92
  -Thermal -1,000,000 Btu/hr Hot Water

- Peak Shave Mode at Engine Speed of 3600 rpm
  -Electrical - 320 kW, 460 V, 3 Ph, 60 Hz, Power Factor 0.94
  -Thermal - 2,100,000 Btu/hr Hot Water*

Input
- Cogeneration Mode at Engine Speed of 1800 rpm
  -1830 SCFH Natural Gas (1020 Btu/scf HHV)
  -Coolant Pump Motor 1.2 kW, 460 V, 3 Ph, 60 Hz
  -Controls and Misc -1 kW, 460 V, 3 Ph, 60 Hz

- Peak Shave Mode at Engine Speed of 3600 rpm
  -4130 SCFH Natural Gas (1020 Btu/scf HHV)
  -Coolant Pump Motor - 4.7 kW, 460 V, 3 Ph, 60 Hz
  -Controls and Misc -1 kW 460 V, 3 Ph, 60 Hz

Efficiency
- Cogeneration Mode
  -Electrical - 29.2%
  -Combined Electrical and Thermal - 82.8%

- Peak Shave Mode
  -Electrical - 26%

Controls
- Microprocessor based with Tecogen Single Board Controller. Fully automatic start-up, monitoring, load following, dual speed and power operation and shutdown. Programmable set points. Digital display. Optional remote monitoring.

Dimensions     12' long × 7 6" wide × 6'6" high

Weight 10,000 lb approximately

*Based on load return water temperature of 170°F
Specifications subject to change without notice.

*Chapter 17*

# Interface Between Private and Public Entities in the Cogeneration Industry

*Chapter 17*

# Interface between Private And Public Entities in the Cogeneration Industry

*David S. Milne, Jr.*
*Gas Energy Inc.*

C ontract abrogation controversies relating to PURPA have been brought on by changes in the energy industry. IPPs have responded with alternative power purchase agreements. In light of these changes, an alternative to standard IPP contracts is power sales contracts negotiated between unconventional parties such as public/private enterprise undertakings. This chapter explores these developments.

Changing attitudes of federal and local governments forecast deregulation in the energy industry and the end of PURPA. PURPA, intended to encourage the development of alternative sources of power, had specific requirements for contracts between cogenerators, hosts, and utilities which facilitated the financing of those projects. The recent abundance of low-cost energy, and the threatened repeal of PURPA, has forced IPPs to seek out different financing arrangements.

In the 90's, when energy became available at costs well below the long-run avoided cost contracts, utilities which had signed with IPPs began to challenge the contracts, demanding abrogation.

The Federal Energy Regulatory Commission (FERC) has strongly supported existing long-term contracts made between IPPs and utilities. In cases where utilities did not contest the pricing structure at the time of contract, FERC denied utilities' claims that avoided costs agreed to at that time should be renegotiable if those costs decrease in later years. However, utilities are pursuing their cases on higher appellate levels. With the federal government seriously considering deregulation, with the future of PURPA questionable, and with energy producers readily available, old IPP/host/utility energy models are no longer viable.

This situation is exemplified by the case which the New York State Electric and Gas Corporation (NYSEG) filed against Lockport Energy Partners, to be discussed later in this article.

# PROJECTED
# FUTURE ENERGY ENVIRONMENT

### Merchant Plant
One proposed model for the new deregulated environment is the merchant plant. This facility is built on the speculation that there will be utility customers or end users for its energy after it is built. Such plants would probably have requirements contracts with multiple utility buyers, and would look for locations with many transmission opportunities.

The basic difference between the new merchant plant model and the historical Cogen/host/utility model, is the financing method. Cogen models used project financing to garner financing credit based on the contracts for LRACs and capacity payments. Comparatively, the merchant plant model, with no such set contracts, will require its developers to use corporate financing and commit a much higher percentage of capital.

### "Specialty Situations" and Corresponding Contract Structure
Specialty situations describe needs, other than low cost power, which cogen plants are able to fulfill for their buyers. The more of a buyer's needs an IPP can meet, the more attractive it becomes. Because future agreements with buyers are likely to be based on requirements, selling multiple services or meeting multiple buyer needs are important strategies for IPPs looking to satisfy financial objectives.

### Specialty Situations
There are many types of specialty situations. Often, buyers who need to invest limited money in infrastructure projects look for IPPs

which will assume economic responsibility for renovating and/or maintaining buyers' plants as part of requirements contract agreements.

Buyers requiring plant upgrades for increased efficiency or those seeking to bring their plants into alignment with environmental standards, have another special need which can similarly be met by plant privatization. By turning over the responsibility of plant maintenance to IPPs, buyers are not only freed from technical and equipment costs but are also spared the cost and responsibility for human resources required to run the plant.

Buyers, who for "control reasons," need a guaranteed, uninterrupted source of power to protect against costly disruptions, can meet this special need with IPPs. Buyers requiring "predictably priced power," will reach agreements with IPPs for rates based on a formula price and a generally available index. This serves as an attractive alternative to the common pricing method derived from rates set by the Public Service Commission.

# CONTRACT ABROGATION CONTROVERSY

General Motors' Lockport plant is an example of changing attitudes. The plant was designed and built with complete system redundancy, to satisfy General Motors' requirement that a cogeneration plant be reliable and available to supply the needs of GM's Harrison Division located in Lockport, New York, 100 percent of the time. The expected availability for the plant, including scheduled and unscheduled maintenance, is in excess of 95 percent. The plant supplies steam and electricity to a General Motors facility that manufactures automobile air conditioners. Excess electricity is sold to New York State Electric and Gas Corporation (NYSEG).

The $209 million project, which began commercial operation in December 1992, required the construction of a nominal 174 megawatt natural gas-fired combined cycle cogeneration facility. The major components of the project include:

- Three General Electric Frame 6 gas turbine generators
- One General Electric steam turbine generator
- Three heat-recovery steam generators
- An auxiliary boiler

# LOCKPORT HISTORY

### Background

Lockport Energy Associates (LEA) owns and operates a nominal 174 MW cogeneration facility in Lockport, New York. LEA has a power purchase agreement with NYSEG, executed on March 26,1990, pursuant to which NYSEG purchases the facility's net electric output (less any power sold to General Motors, Harrison Division).

The payment rates under the Lockport agreements are fixed for the terms of the agreement. The rates were discounted off of estimates of NYSEG's long-run avoided costs (LRACs). NYSEG's LRAC estimates were calculated by the New York State Public Service Commission (New York Commission) in 1988.

In its petition to FERC, NYSEG complained that the New York Commission required NYSEG to enter into the agreements with Lockport despite NYSEG's concerns that the agreements did not adequately protect NYSEG's ratepayers against the risk of payments in excess of avoided costs. Relying on the projections of two independent analysts, NYSEG argued that the rates under the agreements were unauthorized under PURPA because they will exceed NYSEG's avoided costs throughout the terms of those agreements.

### FERC Ruling

FERC affirmed the validity of the contract between NYSEG and Lockport, and refused NYSEG's request to alter the contracted financial arrangements between the two companies, based on PURPA, the Commission's regulations or public policy.

Reasoning: "The contracts at issue allocated risks to both the purchaser and the sellers... QFs bear development risks not experienced to the same extent by traditional utilities. As a result, they must rely on their power purchase agreements to obtain project financing, and we have recognized the importance of contractual reliance for this purpose. If we were to grant the relief requested by NYSEG and allow the reopening of contracts that had not been challenged at the time of execution, financeability of such projects would be severely hampered. Such a result is not, in our opinion, consistent with Congress' directive that we encourage the development of QFs."

# CHANGING ATTITUDE OF NY COMMISSION

The New York State Public Service Commission did not forecast the amount of power produced by Ipps. According to IPPNY, 18.3% of the power sold to New York State customers in 1994 was produced by independent power producers. By strongly enforcing PURPA, FERC encouraged development of cogeneration projects, but never expected so many projects to come to fruition. The abundance of plants has resulted in a ready supply of energy, increasing competition and lowering prices.

**Utilities which seek relief for ratepayers through reduction in IPP purchases are ignoring the source of increased costs.**

Because of the costs involved in upgrading and maintaining energy facilities, and the lack of designated funds for these purposes, public utilities' energy facilities are inefficient and costly to run. Although utilities balk at the prices charged by IPPs, it is often cheaper for utilities to purchase energy from IPPs than to create it using obsolete facilities.

Costs incurred by governmental tax increases are generally passed along directly to the consumer, relieving utilities of the tax increase burden. This off-set burden is another argument utilities have used for lowering IPP prices.

The New York Public Service Commission intervened with FERC on behalf of NYSEG.

NY Commission asked FERC to relieve NYSEG of contract obligations based on LRACs calculated by the Commission. NY Commission said, "Because avoided cost estimates for NYSEG and other New York utilities have dropped since 1990, the purchase prices contained in the Lockport agreements are a burden on NYSEG's ratepayers and have contributed to economic depression in NYSEG's service territory."

A capacity glut in the northeast threatens the economic health of certain New York State utilities. Idle utility-owned capacity will hold down marginal electric costs for a long time.

# ALTERNATIVE POWER PURCHASE AGREEMENTS

Changing attitudes promote alternative power purchase agreements. With the demise of fixed, long-run avoided cost contracts, IPPs in the future will depend on selling energy to hosts and others based solely on hosts' needs or requirements.

The contract structure will be dependent on the host. The viability of requirements contracts is based on the dependability of the through-put requirements, or a host's capacity to purchase enough power to meet an IPP's financial requirements. The host in this situation has become the IPP's main credit support, whereas the host was previously only needed for the IPP to qualify as a cogen facility under PURPA.

When the IPP was used for qualification purposes, the utility was the main credit support for funding. Since the host is now the main credit support, IPPs carefully examine a host's business to determine long-term viability and credit risks, as opposed to the more limited review made when the hosts' importance was primarily related to QF issues.

Requirements contracts will be dependent on product throughput rather than capacity payments. Because requirements contracts are dependent on the actual energy used, IPPs enhance their ability to attract customers and ensure sales by offering multiple services and meeting many hosts' utility and other needs. Offering thermal as well as electrical energy, structuring deals to renovate and maintain hosts' physical plants, and producing high-quality uninterrupted energy sources for hosts, are examples of specialty situations IPPs use to negotiate long-term requirements contracts with hosts.

# JOHN F. KENNEDY INTERNATIONAL AIRPORT

In 1989, the Port Authority of New York and New Jersey needed to expand and upgrade John F. Kennedy International Airport's heating and cooling facilities. All major system components required improvements, including the existing central heating and refrigeration plant and thermal distribution system.

The Port Authority turned to private investors—GEI and its associates, known as KIAC Partners—for the solution. The need for capital investment was a primary factor driving the Port Authority's choice. KIAC assumed all central utility infrastructure investment requirements.

GEI constructed a 100-megawatt natural gas cogeneration plant which supplies electricity and thermal energy, in the form of hot and chilled water, for heating and cooling the airport's central terminal area. Incidental electricity is sold to Con Edison, the local utility. GEI renovated and expanded the Central Heating and Refrigeration Plant (CHRP) to increase the cooling capacity to 28,000 tons and the heating capacity to 225 million Btu/hr. Renovation and expansion of the Thermal Distribution System (TDS), was also required.

### KIAC's Power Purchase Agreement

KIAC negotiated a requirements contract with JFK in which the airport agreed to purchase most of KIAC's electrical and thermal output up to the needs of the airport.

Thermal energy is a high-value product for JFK which is used to secure passengers' safety and comfort. In selling thermal (hot and chilled water) and electrical energy to JFK, KIAC provides for the airport's special needs from a single source and guarantees itself a customer for three products. Selling multiple products meets IPPs' financial objectives formerly met by high LRACs and capacity payments.

The unused energy is sold to Con Ed at a short-run avoided cost when that cost is competitive for KIAC. While a competitive cost is economically beneficial for KIAC, it is also technically advantageous. By servicing Con Ed's needs, KIAC simultaneously receives income and keeps its turbines running at full load to maintain optimal mechanical efficiency. Through the sales at the short-run avoided cost tariff, KIAC dispatches itself based on economic and technical considerations.

The transactions were driven by cost objectives other than securing low electric rates. By assuming economic and operations responsibility for renovating, maintaining and running the physical plant, KIAC enables Port Authority to dedicate its capital to core business needs (i.e. bridges, tunnels, and airport infrastructure repairs). In order to supply JFK with thermal energy, KIAC rehabilitated JFK's central heating and refrigeration plant while building the cogeneration plant.

# WHAT IS THE LIKELIHOOD OF AN INTERFACE BETWEEN PRIVATE AND PUBLIC ENTITIES IN THE COGENERATION INDUSTRY?

### Privatization is Popular on both Federal and Local Levels

President Clinton's administration has been working feverishly to develop a privatized model for Medicare. New York City Mayor

Rudolph Giuliani, ordered a study to explore the benefits of selling Kennedy and LaGuardia airports to private entities. As privatization increases and entities look for ways to cut costs, public and private institutions such as airports, universities, hospitals, and manufacturers, become excellent candidates for "inside the fence" cogeneration facilities.

The increased governmental interest in privatization is also opening doors to retail wheeling. When retail wheeling comes into play, energy suppliers will have access to transmission lines, and will be able to serve multiple retail buyers in different geographical areas. This expanded access will improve generators' financing opportunities, and encourage competitive pricing in the industry.

What are the mutual benefits of long-run requirements agreements? Public agencies are ideal partners. In a deregulated environment, an IPP seeking an economically feasible requirements agreement wants its potential host to demonstrate a stable, long-term need for energy. Through operations data accumulated during a public agency's lifetime, IPPs have solid information which can be used to evaluate the agency's current power needs and estimate the potential for increased needs.

Public agencies are likely to exist for a long period of time. They generally pay their bills, despite public spending crises.

**IPPs Offer Advantages to Public Agencies.** Relieved of the need to invest financial resources in their energy plants, public agencies can apply those resources to other infrastructure needs. Plus, IPPs provide low cost, dependable energy.

The requirements contract with the appropriate customer offers a practical solution to the financing hole opened by PURPA's questionable future. In particular, the private/public combination permits contract structures which, although greatly modified from the historical standard, will enable IPPs to operate in an area with future growth potential.

Changes in the energy industry and shifts in government policy have inspired IPPs such as GEI to generate creative solutions for the future. By responding to hosts' needs rather than rejecting them as demands, IPPs may find the means to secure long-term contracts and meet their own financial objectives.

*Chapter 18*

# Using In-House Expertise
# In Negotiating Power Sales Contracts
# For Industrial Cogeneration Plants

*Chapter 18*

# Using In-House Expertise In Negotiating Power Sales Contracts for Industrial Cogeneration Plants

*Roger Yott, P.E.*
*Air Products and Chemicals, Inc.*

<br/>

nergy has always been a strategic component of Air Products and Chemicals production costs. In fact, Air Products is among the top consumers of electricity and natural gas in the U.S. Consequently, Air Products has developed a multifaceted Corporate Energy Department. The advent of PURPA in 1978 and the success enjoyed by Air Products in selling industrial gases "over the fence" to industrial customers as an integral part of their manufacturing system led Air Products into the industrial cogeneration business.

This chapter briefly summarizes Air Products' entry into the industrial cogeneration market and the role that Air Products Energy Department has played in making this new business focus a success. It highlights how Air Products has been able to transfer its in-house expertise in purchasing power to the marketing, bidding, contract negotiation and avoided cost forecasting functions so critical in the successful development of industrial cogeneration opportunities. At Air Products we believe our long association with the utility industry first as a cost-conscious customer and more recently as an electric energy supplier has enhanced our competitive position. The same success story could be repeated at your company if you know what to look for and are not afraid to expand the horizons and responsibilities of your energy department.

Air Products and Chemicals has gone from being one of the nation's largest consumers of electricity to one of the leaders in power production. This article reports how the expertise developed through buying from utilities was transferred to the selling side. Several key areas affected by this expertise transfer are:

- Market Analysis
- Bid Preparation
- Contract Negotiation
- Avoided Cost Forecasting
- Public Policy Shaping

# ABOUT AIR PRODUCTS

Since its founding in 1940, Air Products has grown into a company with almost $3.0 billion in sales and net income of over $249 million. Capital expenditures in FY 1991 totaled $657 million adding to the more than 150 plants and 5 research centers now in operation in 24 countries. The company employs over 14,000 employees worldwide with 4,000 located at its corporate headquarters in Allentown, Pennsylvania.

# ENERGY AREA ORGANIZATION

Over the past 15 years, price volatility of energy has been substantial relative to other basic commodities. In order to manage this variability and to minimize the negative impact of unexpected changes in the supply and price of such critical components of its business, Air Products established one of the country's most comprehensive energy management teams.

In 1983, Air Products became a marketer of energy and energy services through a growing involvement in power production at our affiliated environmental and energy businesses. Our customers and partners benefit from the same experience and expertise that has supported the operations of our internal businesses.

The company's energy management team has experience in all aspects of energy management from fuel procurement to price forecasting to contract negotiation. The organization chart shows the breakdown of the three major functions within the department:

- Electricity Supply and Energy Policy
- Power Generation and Marketing
- Primary Energy and Energy Economics

### Energy Area Organization Chart

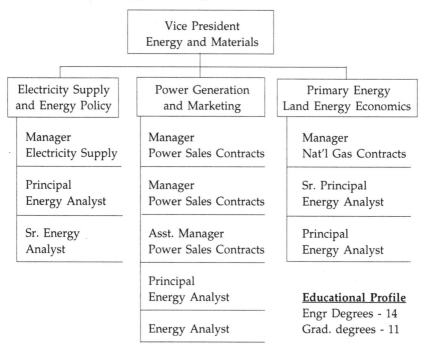

## ENERGY MISSION

Since its founding in 1978, the energy area has been responsible for the worldwide supply of the company's energy needs, but with the company's new ventures, our mission has been expanded to include providing functional support to those businesses that produce energy for sale or other energy-related services. The remaining five items span all three functional areas whether buying or selling electricity or gas.

- Ensure compliance with energy laws and regulations
- Identify factors impacting plant siting
- Optimize productivity gains
- Formulate energy policy
- Direct energy litigation

In some cases, the diverse needs of the corporation conflict with each other particularly where we are buying and selling to the same utility.

**Energy Area Mission:**

---

- Provide worldwide supply of company's energy needs
- Furnish functional support to business area in production and sale of energy and energy-related products/services
- Ensure compliance with energy laws and regulations
- Identify and analyze energy factors impacting new plant locations
- Optimize energy productivity gains
- Formulate and promulgate corporate energy policy
- Direct energy litigation

---

# ENERGY WHEEL

Another way of looking at the diverse responsibilities that exist in the energy management team, is through the energy wheel.

This wheel shows the various functions that the group manages. Some, like public policy shaping and contract negotiation, span all three areas while others like supplier evaluation apply only to the "buying" side of the group.

The addition of the power generation side of the business grew from the expertise that existed in the department through its purchasing experiences. Basically, the new businesses that the company was getting into provided the department with new opportunities for growth based on our realization that there was another side to negotiating with our longtime utility suppliers.

**Energy Area Functions**

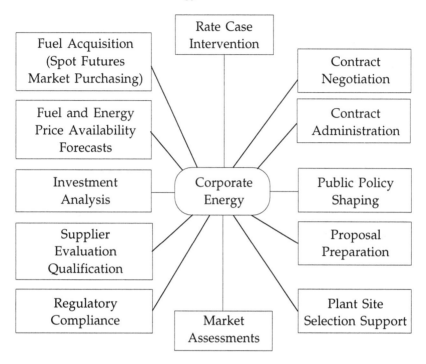

# REF-FUEL WASTE-TO-ENERGY

In 1983, the company began to look at ways to use the on-site capability that it had established through its industrial gas and chemical businesses to enter into the growing field of energy production and in particular, waste-to-energy plants.

American Ref-Fuel was formed as a joint undertaking with Browning Ferris Industries of Houston, Texas, a leader in the waste processing field, to build, own, and operate solid waste-to-energy conversion projects. With Air Products operating expertise and BFI's knowledge of the waste disposal industry, the joint venture looked like a winner.

**REF-Fuel List**

| Location | Size (TPD) | Size (MW) | In Service |
|----------|-----------|-----------|------------|
| Hempstead, NY | 2,250 | 65 | 1989 |
| Essex County, NJ | 2,250 | 65 | 1990 |
| Preston, CT | 600 | 15 | 1992 |

# THE REF-FUEL LIST

This list shows the Ref-Fuel plants presently operating. Our first plant was a $400 million state-of-the-art facility in Hempstead, New York on Long Island. This plant burns up to 2,250 tons per day of municipal solid waste to generate over 70 MW of electricity. The Essex plant in New Jersey was completed in 1991. The 15 MW Southeast Connecticut project was completed this year.

# ENERGY SYSTEMS

The success of the waste-to-energy venture with BFI encouraged the company to look further at the possibility of entering into the independent power production industry. As part of the energy efficiency programs at our plants, facilities that provided the right steam and power loads were studied for cogeneration opportunities. In 1985, our New Orleans chemical facility was the first to install a 20 MW gas-fired unit to supply all of the facility's steam and most of its electricity needs.

Our experience of having negotiated the first Ref-Fuel power sales contract helped us immensely when we first approached New Orleans Public Service with the fact that we were installing a cogen plant and needed supplemental power, backup power, and an interconnection agreement.

# EED COGEN PROJECTS

Additional cogen facilities were put in at our Pasadena and Rozenburg chemical plants and all of our other plants have been and continue to be analyzed. In most cases, either a sufficiently large steam load does not exist or the plants are being run using low cost, interruptible electricity which is not economical to displace with cogeneration.

### List of EED Cogen Projects

| Location | Thermal Host | Technology | Size (MW) | In Service |
|---|---|---|---|---|
| New Orleans, LA | APCI | GAS-CT | 65 | 1984 |
| Pasadena, TX | APCI | GAS-CT | 65 | 1985 |
| Rotterdam, Netherlands | APCI | GAS-CT | 15 | 1987 |
| Stockton, CA | COM PRODS | Coal-Fluid Bed | 65 | 1988 |
| Cambria Co., PA | SPP | Waste Coal-Fluid Bed | 65 | 1991 |
| Orlando, FL | APCI | GAS-CC | 15 | 1994 |

With our success with Ref-Fuel and at our existing facilities, it was decided in 1986 to pursue the third party cogeneration field but concentrating on what we did best, that is, to engineer, build, and operate complex systems for our customers. Air Products decided to concentrate on solid-fuel opportunities where our on-site experience could be best utilized.

Our Stockton, CA, plant serving PG&E and Corn Products came on-line in 1988 and has achieved an on-stream factor of 95% with almost 100% during the summer on-peak periods. It utilizes a coal fired fluidized bed boiler.

Our second project utilizes a waste coal fired-fluidized bed boiler. Located in Cambria, PA, it generates 85 MW for sale to Pennsylvania Electric Company. Our third project is a 115 MW combined cycle plant at our Orlando air separation plant selling power to Florida Power Corporation.

## MARKET ASSESSMENTS

One of the primary responsibilities of the power generation marketing group for the various energy businesses in the company is in assessing the markets available for our products and services. In the cogen and waste-to-energy field that means determining which utility is going to need power and when. Our expertise in purchasing from utilities helped us to better understand how they plan to meet their future needs, and helped us to make more informed market assessments.

# TOP TEN STATES

In addition to assessing the need for power nationwide and the regulatory environment that affects who will supply that need, our department assists the business areas by helping them concentrate on specific states and utilities based on their future need for power. Illustration C shows our top ten list of states from several years ago. These would be the states where Air Products concentrates most of its cogeneration marketing efforts.

The list is updated approximately every six months. The qualifications for being on the list include, obviously, utilities that are planning on coming out with requests for proposals or RFPs to purchase blocks of power, states that do not have an RFP process but have an obvious need for future power resources, states with high retail rates and states that would be amenable to solid fuel power generation.

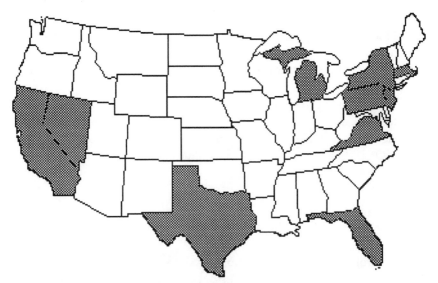

**Top Ten States for New Non-Utility Generation, 1989--1999**

# PROPOSAL PREPARATION

With more and more utilities contracting to purchase new power through an RFP process, our department works very closely with the business areas in the preparation of bids. This is becoming a tremendous task. The ability to successfully bid may indeed be what separates the real players in independent power from the smaller developers.

Preparing a bid proposal is time-consuming and expensive. With major RFPs coming out at the rate of more than one per month, manpower and cost become important considerations. Often, we decide not to bid the more marginal RFPs in order to put more effort into those we feel we are more likely to win. Our expertise in purchasing from utilities has helped us in several tangible ways where we have submitted bids.

# TRENDS IN CAPACITY ACQUISITION

A number of utilities are moving towards competitive bidding to acquire their future resources. We feel that, because we are generally known by the utilities through our existing energy activities, we come into the bid with "instant" credibility. Also, from our long association with the utilities we speak utility-ease fluently allowing us to communicate effectively with the utility.

Finally, having worked with utilities, we are better able to identify areas where they have real concerns that must be addressed to successfully close the deal. In short, our purchasing expertise allows us to develop effective bid proposals.

We are able to identify and concentrate on developing projects in those geographic areas of most benefit to the utility; supply them the information clearly and concisely needed to evaluate our proposal, and to offer to provide them the resource operating flexibility they want at a competitive price. (See box on following page.)

# TRENDS IN UTILITY CAPACITY ACQUISITION PROGRAMS

Utilities are beginning to weigh the <u>bidder</u> as much as the <u>bid</u>. Though winner failure rate has not been excessive, many projects are saved by selling out to more substantial entities that complete the project.

---

### Summary of Competitive Bidding (as of November 1991)

| | |
|---|---|
| # of RFPs issued: | 83 |
| # of MWs requested: | 21,141+** |
| # of Projects/MW bid total:* | 2,334/166,752 |
| # of Projects/MW winning:* | 339/12,839 |
| # of Projects/MW cancelled:* | 33/1,518 |
| # of Projects/MW on-line:* | 69/1,979 |
| # of States/Provinces where bidding used: | 26 |
| # of Utilities issuing RFPs: | 56 |

*MWs rounded
**No limit on 3 RFPs

---

Unfortunately in many cases the original bidder won because they made claims in their proposal that could not be met so additional negotiations must take place. Utilities are moving more towards the traditional bidding systems we are all familiar with in purchasing equipment and material.

### Trends in Utility Capacity Acquisition Programs

• Utilities want the power contract to be the last document required for financial closing.

• Utilities want to deal with only those substantial entities able to complete the development of the project.

• Utilities want to award capacity which results in the greatest value to its ratepayers.

• Capacity acquisition programs will see more, not less, competition in procurement of new generating capacity.

More and more analysis is being done by utilities of the least cost method of providing for their capacity needs. Many states have hearings going on right now weighing all of the various methods of meeting their energy needs through the coming decade. Our experience with purchasing electricity and participating in various utility commission rate proceedings has been translated into successful participation in planning proceedings.

# SUCCESS FACTORS

The success of Air Products' energy businesses is related to the company size. It is large enough to be able to finance, execute and operate these plants costing up to $400 million to build. Also, the company has been in the business of providing reliable supply of product "across the fence" under long-term agreements for many years. Our extensive engineering capability is used in value engineering and process optimization. And of course, the strength of our energy group in negotiating with our utility customers is the greatest asset.

**Success Factors**

| Traditional Strengths |
| :-- |
| • Finance, execute, operate |
| • Large projects/long term |
| • Value engineering |
| • Process Optimization |
| • Utility power contracts |

# CONTRACT NEGOTIATION

One of the most important functions of our group in support of the company's energy-producing businesses is negotiating the power sales agreement with the electric utility. Much of the expertise for this came from buying electricity for our energy-intensive industrial gas plants.

In fact, many of the first interruptible and time-of-use type power supply arrangements with electric utilities were negotiated by Air Products back in the seventies. Very similar contract issues were encountered since interrupting an industrial plant as a utility approaches its system peak or running a plant based on a utility's real time costs are closely related to supplying a utility with power. Our extensive experience negotiating "low" rates for our industrial facilities has honed our negotiation skills.

As in the case of bidding, keys to success in negotiating lie in having credibility with the utility:

* Speaking their language.
* Understanding the utility perspective to better craft compromise.
* Instant rapport with the utility—you've experienced similar concerns elsewhere which can lead to proposing solutions tried elsewhere—the utility can be more willing to listen to your side.

# POWER SALES AGREEMENT

A power sales agreement is like a puzzle, you have to get all the pieces to fit together into a coherent document. This process can take months or even years to complete. However, we found that our group had the background and experience necessary to negotiate contracts that met both ours and the utilities basic needs without heavy reliance on outside counsel or consultants to negotiate for us.

Our purchase of power from utilities under innovative supply agreements is more complicated than just buying power under a published rate schedule but negotiating supply agreements has provided a strong expertise base for contract negotiation.

# FUEL AND ENERGY PRICE FORECASTS

Energy price forecasting is another important role our group fills.

# PURPA

Almost 15 years after the simple statement in PURPA that a utility must purchase power from a qualified facility at that utility's avoided costs, the method of buying that power and the price to be paid for it continues to be debated. With bidding becoming more and more prevalent, avoided cost is generally being defined as the cost of the highest winning bidder with the utility's own avoided cost used as the ceiling price.

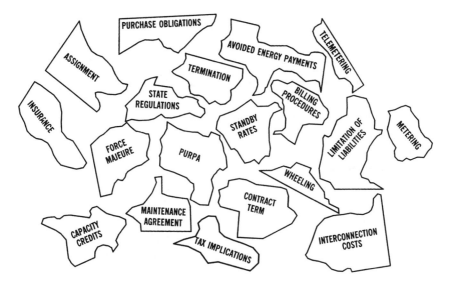

**Power Sales Agreement**

## PURPA—The Basis for Negotiating QF Contracts

- PURPA Section 210 requires the purchase of power from qualifying facilities at avoided costs

  ...The term 'incremental cost of alternative electric energy' (i.e., avoided cost—author's note) means, with respect to electric energy purchased from a qualifying cogenerator or qualifying small power producer, the cost to the electric utility of the electric energy which, but <u>for the purchase from such cogenerator or small power producer, such utility would generate or purchase from another source.</u>

- States with bidding programs have generally defined the "actual" avoided cost for PURPA compliance to be the bid cost of the highest winning bidder, with the utility's own avoided cost used as a ceiling price.

# DETERMINATION OF AVOIDED COSTS

One of the other key roles played by our group both in negotiating contracts and in bidding is in forecasting a utility's avoided costs. Bidding has actually made the job somewhat easier in that usually an RFP has a surrogate generating plant that the bidders use as a benchmark to bid against.

In many of our past contracts, payments from the utility were based on the utility's actual or forecasted avoided costs. As a purchaser of electric energy we've spent hundreds of hours over many years studying how utilities forecast their future rate through our active intervention in rate cases all over the U.S.

This intensive hands on effort has forced us to develop tools to help in the process. These same tools can be used in the critical area of forecasting future utility avoided costs the revenue stream for 50-60% of waste to energy and 90-95% of cogeneration projects.

To help forecast what the utility should be offering us, we worked with an outside consultant to develop a probabilistic computer model to determine a utility's incremental cost of energy production. Although less accurate than a "PROMOD"-type model, it allows us to ball-park a forecast and run sensitivities.

### Determination of Future Avoided Costs

| |
| --- |
| 1. Future prices of fuel |
| 2. Mix of generating units |
| 3. Projected load growth |
| 4. Future operating efficiency |
| 5. Availability of purchased power |
| 6. Aggregate cogeneration capacity |
| 7. General inflation |
| 8. Legislative/regulatory trends |

The key items affecting the validity of our forecasts are the future prices of fuel, generating mix, load growth, and the combined effect of all three. Although recent run-ups in oil prices would have made us think of increased energy revenues, in one of our facilities, the utility's load is down so much due to the economic downturn and recent weather conditions in the Northeast that the marginal generation is mainly coal rather than oil.

The results are avoided costs well below what our forecast would have indicated using normal weather and load growth assumptions.

# PUBLIC POLICY SHAPING

How do we deal with the many laws and regulations affecting non-utility generation on both a state and national level? How do we know that the next legislation coming down the track doesn't shut down one of our existing plants or curtail future activity? How do we best provide our input into the complex world of energy legislation and regulation?

One of the ways we do it is through our state and federal lobbyists as well as our groups' active participation in selected state regulatory proceedings. Our group is responsible for deciding when to bring the lobbyist in and to make sure they understand the issues.

In our company, the lobbyist is a lobbyist for a chemical and industrial gas company, so there is an education process on energy issues. For example, we were able to get waste added to the Solar, Wind and Geothermal Act as well as eliminate the size restrictions. To accomplish this we worked very closely with our Washington office and were able to explain the issues to government legislators through our lobbyists.

## Public Policy Shaping

| |
|---|
| • Legislative and regulatory lobbying |
| • Federal and state level |
| • Independent power issues: |
| —The Clean Air Act |
| —PUHCA reform |
| —Competitive bidding |
| —Transmission access |
| |
| • Intervention groups |
| —National Independent Energy Producers (NIEP) |
| —Independent Energy Producers (IEP) |
| —Independent Power Producers of NY (IPPNY) |
| —Electricity Consumers Resource Council (ELCON) |
| —Industrial Energy Group (IEG) |

New issues in independent power production are emerging every day:

* The Clean Air Act
* PUHCA Reform
* Competitive Bidding
* Transmission Access

No one company can get involved in all the issues by itself—particularly companies like us that have other businesses with other interests. One of the key ways to "keep in touch" is through organizations that promote your needs. We belong to some 20 organizations just related to energy covering both the consuming side of the company as well as the producing side. This allows us to leverage our own internal skills, increasing our ability to favorably shape those public policies that affect our long term competitive position in the non-utility generation industry.

## SUMMARY

In the area of power production, the buying and selling sides are becoming a little more fuzzy, with no bright line between supplier and customer. Although we may put in our own generation, we will still need backup power and the utility should be justly compensated.

On the other hand, not every interconnection has to be a work of art costing millions of dollars to install. This article portrays how Air Products got involved in the business of power production and how we have used our internal expertise rather than always going outside to deal with our utility customers/suppliers.

Whether you are considering your own cogen plant or bringing in a third party developer, it pays to understand the world of non-utility generation, at least from a general point of view. There are plenty of consultants out there ready to provide you with help, but at a large price tag. As you get more involved in cogen, you should seriously consider developing the internal expertise to deal with the many challenges this business has to offer. It may just be the difference between success and failure.

*Chapter 19*

# Post-Contract "Neo-Gotiation": How Cogenerators Can Sell Utilities

# Post-Contract "Neo-Gotiation": How Cogenerators Can Sell Utilities

*Paul Gerst, P.E.*
*Enercology Associations, Ltd.*

---

Negotiation in the classic sense envisions a roomful of people hammering out details of a difficult contract for a proposed event. Things have changed subtly in the cogeneration industry.

When cogeneration's federal mandate was first exercised, the standard utility posture called for dragging out contract negotiations interminably until the cogenerator gave up. California's standard offers were one answer to this particular dilemma.

A decade later, there seems to be a new watchword, "neo-gotiation." While the initial power purchase agreement may have been signed by both parties, the contract's detailed implementation and its administration is still another matter, and that is the element that is subject to "neo-gotiations."

Today, many of our post-contract "neo-gotiation" situations are a result of bidding procedures. Whether they are multi-attribute bids, or whether they are simply price bids, or a second price auction, each one of those particular contracts has a variety of elements in it which are essentially judgmental. Attorneys abhor this concept of a judgmental clause in a contract, but others should view those clauses kindly because they may be the elements which help pay the rent!

The idea of "neo-gotiation" cries out for a sales plan. *First,* there are areas of negotiation which are still open despite the fact that you have a signed power purchase agreement.

*Second,* in those areas of negotiation, there are many important people who are very difficult to get to; nonetheless, these people will be making decisions concerning you and your project.

*Finally,* review for a moment the traditional sales program: There were a couple of fellows in the early fifties who made their living telling just one thing, "Sell the sizzle, not the steak." It may not seem to be totally applicable to a highly technical contract, but nonetheless, sizzle works.

I suggest you start thinking in terms of sizzle because it is inherently the benefit to the user. How does the customer, in this case, the utility, gain from your proposal?

# THE DECISION TREE

There are a lot of hurdles to a successful power contract "neogotiation." Figure 19-1 shows what a utility needs and what you can supply as a potential cogenerator. The very first decision on this tree asks whether the utility needs capacity. If the utility needs capacity, its ability to meet that need depends upon whether the utility has funds available, or it needs funds.

This is an important point! The utility may have been literally forced to give you a power purchase agreement by virtue of a state-mandated procedure; but what is the real need of the utility?

It has to buy capacity if it lacks funds, or it has to count on the PUC saving the day through some sort of magnanimous gesture with respect to the rate base. An alternative could be conservation. But let's follow the decision tree down the center core, namely, that the utility must buy capacity.

As the utility gets down into the must-buy category, it has choices: it can elect to buy private power, or it can make an "IFC" purchase. No one knows what an IFC purchase is because I made this acronym up.

It's an Inter-Fraternity Council purchase, and those of you who have been members of the inter-fraternity council at one time or another will recall that's the way things were divvied up when the new freshman class comes in. The Betas get the Thetas, the Sigs get the Pi Phis!

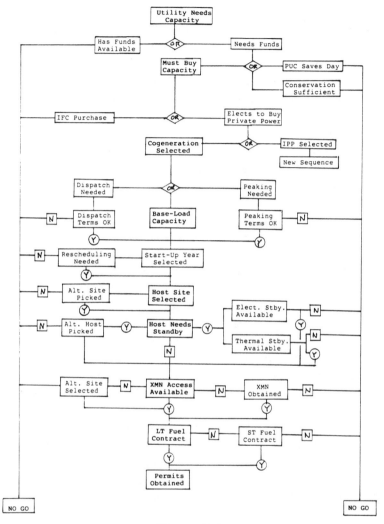

**Figure 19-1.**

An IFC is essentially a purchase from a utility subsidiary. If the utility chooses to purchase power from a private party, it has choices: cogeneration under the PURPA scheme, or an IPP. If it chooses the IPP, a whole new sequence is involved, and because this is cogeneration, we will ignore the IPP choice: Independent Power Producers (IPP's) do not require a thermal host as do PURPA-sanctioned power producers. We will go down the center core presuming the utility selects cogeneration. Each decision, as one comes down the "tree," may have been unresolved in the power purchase agreement. But not all of them will be totally resolved, and possibly none of them is irrefutable, or non-renegotiable! Man's ingenuity seems to be infinite, and there are dozens of new ways to have a "neo-gotiation" on virtually any topic in this article! This tree was patterned after a specific case, and these were the issues in that specific case.

The concept of loading is very important: Dispatchable, baseloaded, or peaking? If an appropriate deal can be made, peaking could be attractive. If the customer needs dispatchability, and if you cannot provide it, you will fall into the side gutter and that goes down to No-Go. You don't have a project anymore. Presuming that baseload capacity is required, the next major hurdle could be the start-up year.

The start-up year is often a major issue. If the start-up schedule could not be met, that would have been another no-go for the project.

The selection of the host site limits the selection of a host. Had the host's existing site been unacceptable for any number of reasons, clearly an alternate site would have been required. If an alternate site were not available, that's another reason to scrub the project.

Once one has a host, one has to deal with the issue of the host requiring standby electric power, and standby thermal power. If standby is needed, it may impose requirements for special rate structures or special concessions, either by the host, by the cogenerator, or perhaps by the utility within the restraints of its rate structure.

Transmission access in another very important consideration. That reveals a whole myriad of issues. For example, presuming that one does not have transmission access at the site selected, can he obtain rights of way? Can he modify the substation at an affordable cost to be able to accept the new incremental power? Transmission access failures have scrubbed many a project.

# THE PERMIT INTERFACE

The point of these considerations, concluding with the long-term fuel contract and the short-term fuel contract, is to illustrate that it is an interdependent negotiation that you enter into, a "neo-gotiation," in which not only the utility is involved but also you—and not just you, the cogenerator who is supplying electric power; but you, the cogenerator who has arranged for the financing; and you, the cogenerator who has arranged for the permitting! A change in almost any one of these items can result in cataclysmic changes in the type of permitting and the type of activity that's necessary for your project to succeed.

The permitting procedure involves a huge number of agencies. Figure 19-2 is a brief list of the types of permits which are usually required for cogeneration projects. As you can see, the list is lengthy!

Envision, if you will, the circumstances in which you are "neo-gotiating" a contract with these permits in place and have to redo them. We permitting engineers look upon that as the Elysian Fields. Developers might think otherwise!

## Figure 19-2. Common Permitting Units.

**City or County**
Local General Plan Amendment
Specific Plan
Zoning Ordinance Amendment
Special or Conditional Use Permit
Subdivision Map Approval

**Air Pollution Control Districts**
Authority to Construct
Permit to Operate

**Coastal Commission**
Coastal Development Permit

**Special Area Conservation and Development Commission**
Development Permit

**Energy Commission**
Notice of Intention and Application for Certification

**Conservation**
Oil, Gas, or Geothermal Well Permit

**Fish and Game**
Stream or Lake Alteration Agreement
Standard Suction Dredging Permit
Special Suction Dredging Permit

*(Continued)*

**Forestry**
  Timber Harvesting Plan
  Timberland Conversion Permit

**Health Services**
  Hazardous Waste Facilities Permit

**Housing and Community Development**
  Division of Codes and Standards
    Permit to Construct

**Parks & Recreation**
  Right-of-Way

**Transportation**
  Encroachment Permit
  Airport and Heliport Permits

**Water Resources**
  Approval of Plans and Specifications and Certificates of
    Approval to Construct or Enlarge a Dam or Reservoir
  Approval of Plans and Specifications and Certificate of
    Approval to Repair or Alter a Dam or Reservoir
  Approval of Plans and Specifications for Removal of a
    Dam or Reservoir.

**Public Utilities Commission**
  Certificate of Public Convenience and Necessity

**Reclamation**
  Encroachment Permit

**Waste Management**
  Solid Waste Facilities Permit

**Lands Commission**
  Dredging Permit
  Negotiated Mineral Extraction Lease
  Geothermal Exploration or Prospecting Permit
  Land Use Lease
  Prospecting Permit

*(Continued)*

**State Water Resources Control Board**
Permit to Appropriate Water
Statement of Water Diversion and Use

**Regional Water Quality Control Board**
National Pollutant Discharge Elimination System
(NPDES) Permit
Waste Discharge Requirements
Toxics Pits Cleanup Act Requirement
Underground Storage of Hazardous Substances Permit

**Regional Planning Agency**
Project Permit

**United States Army Corps of Engineers**
"404" Permit

**United States Department of Interior**
Bureau of Land Management

**United States Department of Agriculture** United States Forest Service

Figure 19-3 is a decision tree giving an idea of the sequence that each permit could involve. You have not only a decision tree, you have a decision tree with decision branches, decision trunks, and interlocking decisions within a forest of other interdependent trees.

I would suggest that a contract "neo-gotiation" is not a matter to be taken lightly. It is a very serious hurdle to success even in this simplified case. Every hurdle that you meet in "neo-gotiation" involves a new sales job.

# PEOPLE ARE MOST IMPORTANT

People are of primary importance to your from a standpoint of selling your project, and in selling the features of your project. It is not so much selling the project as a whole, it's selling features that are necessary to make it go. These are prosaic old sayings which we have all known for all time.

Call them what you wish, but they are essentially the elements which we have always considered whenever we have a problem that needs solving. We ask what is needed, who is the key man, what can we do to fill his needs, when is that need best met, and why is our answer the best one?

That is what I would like to leave with you... the essential idea that "neo-gotiation" is really a mechanism of answering the usual questions, and framing them in the format of your customer's business requirements.

Generalized Development Project
Application Process

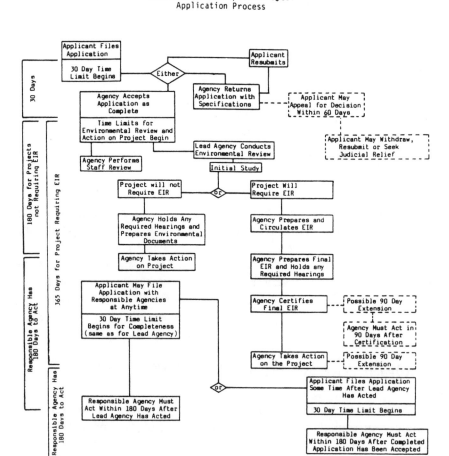

Note: Illustrative only.

**Figure 19-3.**

*Chapter 20*

# A New Market Opportunity: Carbon Dioxide Recovery from Cogeneration

*Chapter 20*

# A New Market Opportunity: Carbon Dioxide Recovery from Cogeneration

*Sam A. Rushing*
*Advanced Cryogenics, Ltd.*

O ver the last few years, the nature of the carbon dioxide industry has changed significantly. New companies have entered independent $CO_2$ production, where cogeneration plants which require QF status, under PURPA regulations, utilize cogenerated steam in an amine-based carbon dioxide recovery process. Such cogeneration projects benefit from maximum efficiencies in steam utilization as provided for with such a steam host; plus new, long term added revenues are created via $CO_2$ production. Such new $CO_2$ plants plus reduce airborne emissions, and supply a long term, stable steam host for a project.

Moreover, the $CO_2$ from cogeneration means of production can create a new carbon dioxide source in strategically important areas where production has always been needed, but not available from any of the traditional means of production.

## CHANGING $CO_2$ PRODUCTION

The carbon dioxide industry has been held hostage by restricted raw feedstock sources, which are often by-product streams from chemical and fertilizer plants, refineries, wells, and pipeline operations. Until recently, most commercial domestic $CO_2$ production has originated from such processes. But these chemical, refinery and pipeline based facilities often are located far away from the local $CO_2$ consuming market.

In addition to logistics, consolidations among refinery and chemical production facilities have further placed hardships on some regional $CO_2$ market demands. However, flue gas based $CO_2$ recovery and production sometimes will establish a new plant in a location where local $CO_2$ production was never available before from other means of production. In most cases, the "raw stream" is purified, and liquefied for commercial usage; in flue gas based cogeneration processes, upstream of purification and liquefaction is the recovery plant.

Therefore, recovering $CO_2$ from cogeneration based flue gas can sometimes serve as the only means of viable commercial $CO_2$ production for a given area or region to be supplied.

# $CO_2$ MARKET

Approximately 5,000,000 tons per year of carbon dioxide are produced domestically, of which some 70% are dedicated to the food and beverage industries. Beyond this official annual production schedule for the United States, there are many additional hundreds of thousands of tons of $CO_2$ per year generated and utilized for internal processes, such as chemical feedstock requirements.

The largest segment of commercial $CO_2$ usage in the U.S. is the food and beverage market. The largest segment of this market is the cryogenic freezing market, which is estimated well over 2,000,000 tons per year in consumption. As for cryogenic freezing, and temperature reduction requirements, $CO_2$ can serve a duel function in fast freezing of a food product, and also create an anaerobic atmosphere in a packaged product, which insures longer shelf life, bacterial reduction, and improved appearance.

Another rapidly growing carbon dioxide market is in PH reduction, and water treatment applications. Specifically, in many processes and industries, $CO_2$ is the safest and most easily controlled moderate (and self eliminating) acid. This serves the pulp/paper, chemical, effluent treatment, and municipal water treatment markets exceptionally well.

$CO_2$ can replace a variety of dangerous mineral acids such as sulfuric, and hydrochloric; it yields harmless carbonate by-products, and serves an "environmentally friendly" purpose in a variety of industries.

Other growing markets which hold great promise are $CO_2$ based enrichment in crop irrigation projects and greenhouse atmospheres. In many test cases, and also in a wide number of commercial projects, $CO_2$ usage in crop irrigation projects and greenhouse atmospheric enrichment applications yields substantial growth, and added volume to a variety of commercial crops. This is a particularly exciting future for the $CO_2$ industry, and new plants which are strategically located.

Carbon dioxide has a substantial future in grain fumigation, for replacement of carcinogenic halogenated hydrocarbon compounds; and a variety of compounds soon to be banned from the grain industry.

Additionally, $CO_2$ usage as a supercritical solvent in the paint, pharmaceutical, and essential oil (extract) industries often serve as a safe and specific solvent.

For the oil and gas industries, carbon dioxide in very large volumes (hundreds of tons per day and beyond) is utilized in enhanced oil recovery projects, (EOR). For these large EOR jobs, $CO_2$ is often delivered without purification. In the event a cogeneration plant which recovered $CO_2$ was within a working distance of an EOR project, much of the plant would in turn be simplified—purification and liquefaction might be eliminated, based upon strategic location from the wellhead.

In some cases, approximately seven thousand cubic feet of $CO_2$ are utilized per (enhanced) barrel of oil recovered. The life of many enhanced recovery projects ranges from 5 to 25 years. With respect to new EOR requirements for $CO_2$, much of this requirement is driven by the cost per barrel of oil, and the relationship to the economics behind improved oil recovery. Certain projects are viable in the world oil economy today, however, much of this EOR work will rise, with improving oil production and pricing considerations in the future.

Other smaller, well stimulation and fracturing requirements for carbon dioxide exist today, and are growing in scope throughout the United States. The stimulation jobs typically use from ten to two hundred tons per job, and require high pressure service equipment. $CO_2$ will always have a place in the oil and gas industries; however a greater share will exist, with improved petroleum economics and production.

# COGENERATION-BASED $CO_2$ PLANTS: COSTS AND REQUIREMENTS

EOR or chemical grade $CO_2$ may or may not meet most specifications considered essential for food and beverage usage, whereas a function of raw $CO_2$ feedstock, or flue gas constituents, the amine-based $CO_2$ recovery plant may produce high purity material, thus specifications close to food and beverage requirements may exist prior to further refinement (assuming such EOR or chemical process facilities use similar plant design). Food grade $CO_2$ is often defined by major food manufacturers, and soon will be by the Compressed Gas Association.

Strict limits on constituents such as $NO_x$, $O_2$, hydrocarbons, carbonyl sulfide, and oils are defined.

What is considered EOR or chemical grade $CO_2$ is generally less refined than a food grade material, and is produced as a gas, not a liquid product. This leads to less capital investment, and lower utility and operator requirements, in such an industrial project.

In terms of utilities and operating requirements for such $CO_2$ from cogeneration plants, steam requirements average 1.5-2.5 tons per ton of $CO_2$ recovered. Power consumption can be 2 MW for a complete 200 TPD $CO_2$ plant, with a $CO_2$ content of at least 8% (V). As for labor and other operating costs, these requirements are minimal where there is a small chemical and inhibitor replacement requirement for the process.

Ultimately, more enriched fuels such as coal will yield a lower capital and operating cost, as a virtue of a greater $CO_2$ feedstock presence. Other specific considerations such as locating an "over-the-fence" market, which would utilize a $CO_2$ gas, or a chemical/EOR grade product, would simplify the process further with less capital equipment needed (as purification/liquefaction). Lower operating requirements, including power and labor, would lead to a variety of means to produce at a lower cost, and serve a simpler or less expensive $CO_2$ market.

# CAPITAL COSTS

As for capital costs, if recovering from typical natural gas fired turbine exhaust at a raw $CO_2$ content of approximately 3.0% (V), the MEA (monoethanolamine) recovery plant may cost from $10 million-$14 million, plus liquefaction and storage facilities. The capital costs vary according to which licensed technology is chosen, which is primarily driven by cogeneration plant fuel type, thus raw $CO_2$ content.

With respect to specific $CO_2$ markets to be supplied, the $CO_2$ plant configuration, and the fuel type which produces raw ($CO_2$ gas): a variety of plant costs, configurations, and produced $CO_2$ cost/ton cases are available for specific projects. Proper investigation of markets, technologies, and costs become essential in making a determination of the $CO_2$ from cogeneration question.

# RELIABLE $CO_2$ FROM COGENERATION PLANTS EXIST TODAY

Two larger, commercial $CO_2$-from-cogeneration plants are currently operating in the United States. Several additional domestic facilities of this type are planned for the future.

The AES Corporation in Shady Point Oklahoma operates a 320 MW cogeneration plant, which recovers and produces 200 TPD of food grade $CO_2$. This is primarily sold to the poultry industry for freezing and chilling requirements.

This plant uses a circulating fluidized bed (CFB) and burns coal, which requires scrubbers to achieve high purity liquid $CO_2$. In order to meet PURPA requirements, about 60,000 pph of steam at 150 psi are consumed in the monoethanolamine (MEA) recovery process. This $CO_2$ plant consumes about 10% of the total power plant parasitic load, or about 11% of the CFB plant output.

The second commercial $CO_2$ from cogeneration plant currently operating in the United States is the Intercontinental Energy corp (IEC) plant in Bellingham, MA. This is a 400 MW natural gas fired cogeneration plant, which also utilizes an MEA based $CO_2$ recovery process for QF status. This plant is approximately 400 MW, and produces approximately 350 TPD of $CO_2$ from the gas turbine exhaust; the product is sold to regional wholesalers. At the Bellingham plant, the licensed $CO_2$ recovery technology is Fluor Daniel's Economine FG process. The two plants are similar. Their QF status requirement is fulfilled by steam utilization in $CO_2$ recovery; however AES utilizes a coal based fuel and IEC a natural gas fuel.

Technologies differ via amine concentration, inhibitor usage, and operating history. The technologies are proven, and have been utilized for many years in many markets outside of the United States, generally to supply captive $CO_2$ requirements, or to produce for the commercial/wholesale markets. The $CO_2$ recovery technologies are well proven, and reliable.

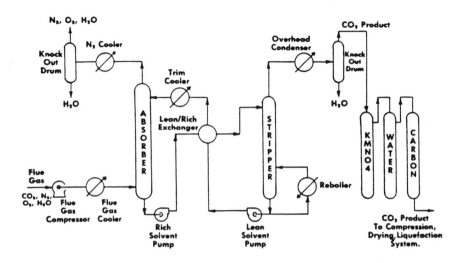

# FUTURE CARBON DIOXIDE
# FROM COGENERATION OPPORTUNITIES

The issue of a steam host has a place, and will be required in certain situations, as related to the cogeneration industry, and PURPA regulations. In addition to fulfilling a primary, or secondary steam host requirement, there are given circumstances where local, or strategically important $CO_2$ production can only be derived from flue gas sources, such as cogeneration exhaust gases.

The carbon dioxide industry has suffered many supply changes in the recent term with respect to the closing or consolidation of the older, once traditional refinery, and chemical process facilities which supplied the raw feedstock. Sometimes the cogeneration-based flue gas feedstock for further $CO_2$ refinement is the only method of making available production in certain areas. Therefore if there is a steam host requirement, and also potential opportunities to recover and produce carbon dioxide to serve a variety of markets, this avenue should be explored.

In some cases, simpler processes and markets can be utilized such as recovering $CO_2$ from cogeneration flue gas and sending the production to an EOR or chemical process requirement. This would take the least capital, and operating requirements.

As for the commercial markets, several tiers of wholesale purchasers may have interest in further refinement; or a take-or-pay purchase of the refined production. Two cases exist with respect to the state in which $CO_2$ could be sold. The product could be sold as a recovered gas from the amine plant, for further purification and liquefaction by a wholesaler; or finished liquid (purified/liquefied) production would be sold to wholesalers or commercial consumers within the local market.

These issues of distributors, direct consuming purchasers, and the sale of a recovered gas or a finished liquid are essential questions to be explored when considering such a facility, and the related markets. It may be best to consider the current majors such as Cardox as a last wholesale purchase alternative, after exploring all other distributor/direct consuming options.

Many $CO_2$ wholesalers who could best handle the wholesale purchase and distribution, or could negotiate the most favorable terms for the cogeneration operator, often do not operate in the region where the plant is being developed; they are looking for opportunities throughout the continent.

Moreover, specific niche markets often exist for best-case purchase of the $CO_2$ production, such as an over-the-fence consuming process, a clustering of industrial processes, or a strong independent regional $CO_2$ distribution network, which can often yield premium pricing per ton, and premium purchase considerations.

The issues of markets, technologies, and operating requirements should be studied in depth during the early phase of planning a $CO_2$ from cogeneration project, and should be in motion for certain secondary steam host requirements, when the primary host is in jeopardy. In other cases, the issue of steam host may be secondary to other considerations, such as special $CO_2$ market opportunities which are driven by plant configurations, fuel types, and regional supply requirements.

*Chapter 21*

# Diurnal Thermal Energy Storage for Cogeneration Applications

# Chapter 21

# Diurnal Thermal
# Energy Storage for
# Cogeneration Applications

*Dr. S. Somasundaram, Daryl R. Brown, Dr. M. Kevin Drost*
*Battelle, Pacific Northwest Laboratory*

## ABSTRACT

Thermal energy storage can help cogeneration meet the challenges of the 1990s by increasing the flexibility and performance of co-generation facilities. Thermal energy storage also allows a cogeneration facility to provide dispatchable electric power while providing a constant thermal load. The first of two studies reported here focused on the relative performance and economic benefits of incorporating a diurnal TES system with a simple-cycle gas turbine cogeneration system to produce dispatchable power during peak- and/or intermediate-demand periods. The results showed that the oil/rock storage system for TES was the most attractive option for the assumed thermal load quality. The second study evaluated the cost of power produced by a combined-cycle cogeneration plant integrated with thermal energy storage (CC/TES/Cogen). The results indicate relatively poor economic prospects for integrating TES with a combined-cycle cogeneration power plant. However, system design optimization and reductions in storage media costs might improve the economics, perhaps enough to make the CC/TES/Cogen an attractive option for incremental peak power production. In addition to the economic considerations, environmental factors due to the use of thermal energy storage with conventional cogeneration systems are also reported.

# INTRODUCTION

The U.S. National Energy Strategy estimates that 200,000 MWe of new electric generating capacity will need to be added between 1993 and 2010. Approximately 40% of the new generating capacity will be for peak or intermediate loads with the remaining 60% providing continuous baseload power generation. Gas turbine schemes such as simple-cycle cogeneration, combined-cycle power plants, and integrated-gasification combined-cycle power plants are becoming the generation options of choice because of their relatively low capital cost, flexibility, reduced environmental impact, and higher thermal efficiency. Thermal energy storage (TES) for utility applications includes a range of thermal energy storage technologies that can further improve the efficiency, flexibility and economics of natural-gas-fired gas turbine options. This is achieved by decoupling power generation from the production of process heat, allowing the production of dispatchable power while fully using the thermal energy available from the gas turbine. The thermal energy from the turbine exhaust can be stored either as sensible heat or as latent heat and used during peak demand periods to produce electric power or process steam/hot water. However, the additional materials and equipment necessary for a TES system will add to the capital costs. Therefore, the economic benefits of adding TES to a conventional cogeneration system would have to outweigh the increased costs of the combined system.

The Pacific Northwest Laboratory (PNL) leads the U.S. Department of Energy's Thermal Energy Storage Program. The program focuses on developing TES for daily cycling (diurnal storage), annual cycling (seasonal storage), and utility applications [utility thermal energy storage (UTES)].

Several of these technologies can be used in a cogeneration facility. This chapter discusses the relative performance and economic benefits of incorporating a diurnal TES system with (1) a simple-cycle, and (2) a combined-cycle gas turbine cogeneration system. The relative benefits of combining a TES system with a cogeneration system were determined by comparing the levelized energy costs of the combined system (for supplying the same preselected steam load) with that of a conventional cogeneration system and a base case electric plant. For example, the combined-cycle study evaluated the cost of power produced by a combined-cycle electric power plant (CC), a combined-cycle cogeneration plant (CC/Cogen), and a combined-cycle cogeneration plant integrated with thermal energy storage (CC/TES/Cogen) systems designed to serve a fixed process steam load. The value of producing electricity was set at the levelized cost for a CC plant, while that of the process steam was for a conventional stand-alone boiler.

# DIURNAL THERMAL ENERGY STORAGE

A number of emerging issues may limit the number of useful applications of cogeneration. One of these is a mismatch between the demand for electricity and thermal energy on a daily basis. Increasingly, utilities are requiring cogenerators to provide dispatchable power, while most industrial thermal loads are relatively constant during a 24-hour period. Diurnal TES can decouple the generation of electricity from the production of thermal energy, allowing the cogeneration facility to supply dispatchable power. Diurnal TES stores thermal energy recovered from the exhaust of the prime mover (gas turbine) to meet daily variations in the demand for electric power and thermal loads.

## Concept

The concept for integrating TES in a natural-gas-fired (simple-cycle) cogeneration facility is shown in Figure 21-1. The facility consists of 1) a gas-turbine prime mover, 2) a heat recovery salt heater, 3) a thermal energy storage system, and 4) a salt-heated steam generator. The gas turbine is operated during peak demand time periods and the exhaust heat is used to heat molten salt in a heat recovery salt heater. Cold salt at 288°C (550°F) is pumped from the cold salt tank, through the heat recovery salt heater, where it is heated to about 510°C (950°F) before being pumped to the hot salt storage tank. Hot salt is continuously removed from the hot salt tank and used as a heat source to meet the constant thermal load. A cogeneration plant with a TES system sized for an 8-hr peak demand period would provide a 30-MWe peaking capacity compared to a similar conventional cogeneration facility that would provide a 10-MWe base-load capacity.

## TES System Description

Depending on the characteristics of the thermal load, a variety of thermal storage systems can be used. Options for thermal storage include:

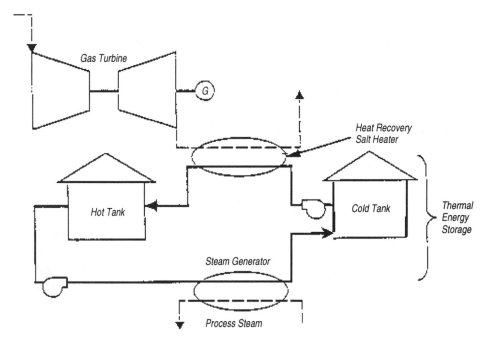

**Figure 21-1.** Schematic of a Simple-cycle Cogeneration Plant with Thermal Energy Storage (SC/TES/Cogen)

- "60/40" Salt TES - The "60/40" salt is an excellent thermal energy storage medium for high-temperature TES applications, using a mixture of sodium nitrate (60 wt%) and potassium nitrate (40 wt%) that can operate at temperatures up to 566°C (1050°F). However, the mixture freezes at 240°C (464°F). To help prevent freezing, these salt systems are usually operated at temperatures above approximately 288°C (550°F). The minimum operating temperature limits the amount of waste heat that can be recovered from a combustion turbine's exhaust because the exhaust can only be cooled to approximately 315°C (600°F). Typically, this type of TES system uses separate hot and cold salt tanks. A more complete discussion of the "60/40" salt TES is presented in Drost et al. (1989).

- Oil/Rock TES - Oil/rock TES is an attractive alternative for inter-mediate-temperature applications. Low-cost heat transfer oils such as Caloria HT-43[b] can operate at temperatures up to 304°C (580°F).

---

[b]Trademark of the Exxon Corporation, Houston, Texas.

The TES system consists of a single large tank that is filled with a mixture of oil and a low-cost filler, such as river rock. The tank is operated to maintain hot oil at the top of the tank and cold oil at the bottom of the tank. This arrangement stratifies the fluid in the tank resulting in minimal mixing between the hot and cold regions of the tank. During normal operation, cold oil is removed from the bottom of the tank, heated in the heat recovery oil heater, and returned to the top of the storage tank. Thermal energy is stored in the mixture of oil and rock. Oil/rock TES is less expensive than molten salt TES, but it is limited to low-temperature applications (< 300°C). Oil/rock TES is described in more detail in Drost et al. (1990).

- Combined Molten Salt and Oil/Rock TES - The advantages of both storage concepts above can be retained by using a combination of molten salt TES for high-temperature and an oil/rock TES for lower-temperature thermal energy storage. This allows the combustion turbine exhaust to be cooled to near ambient conditions, while maintaining higher availability than is possible with oil/rock TES alone.

- Hitec[c] Salt TES - Hitec salt is another molten salt that operates between 454°C and 177°C (850°F and 350°F). It is a mixture of sodium nitrate (7 wt%), potassium nitrate (53 wt%) and sodium nitrite (40 wt%). Hitec salt would allow greater heat recovery from turbine exhaust than the "60/40" salt, but would not be as useful as the "60/40" salt at higher temperatures (> 450°C—as in combined-cycle power production applications). In addition, the Hitec salt is a little more expensive than the "60/40" salt.

Selection of the storage medium will depend on characteristics of the thermal load. If high-temperature thermal energy is required to meet the thermal load, a choice of the "60/40" salt TES, Hitec salt TES, or a combined "60/40" salt and oil/rock TES can be used. Alternatively, if the thermal load uses thermal energy at a temperature below 288°C (550°F), oil/rock TES may be the preferred option.

---

[c]Trademark of the DuPont Corporation, Wilmington, Delaware.

**Benefits**

The use of high-temperature TES in cogeneration applications has the following benefits:

• High-temperature TES will allow a natural-gas-fired cogeneration facility to produce dispatchable power while meeting constant thermal loads.

• High-temperature TES integrated in a natural-gas-fired cogeneration facility allows all power generation to occur during periods of peak demand; the installed capacity of the prime mover will be substantially larger than for a conventional cogeneration facility. A cogeneration plant with a TES system sized for an 8-hr peak demand period would provide 30 MWe of peaking capacity compared to a similar conventional cogeneration facility that would provide 10 MWe of base-load capacity.

• All natural gas is used to fire the combustion turbine (compared to direct natural gas firing of the waste heat steam generator). This results in high-efficiency operation by ensuring that all natural gas is used to produce both electric power and thermal energy.

**Technical Status**

The "60/40" salt TES has been extensively investigated for solar thermal power generation applications. Investigations have included bench-scale testing, detailed design studies, and field demonstrations. Based on the results of these investigations, the Department of Energy and a group of electric utilities are sufficiently confident of the technical feasibility of the concept to embark on the $40 million Solar II demonstration of molten salt central receiver technology. This suggests that the "60/40" salt TES is technically ready for a large-scale cogeneration demonstration. Oil/rock storage has been successfully demonstrated for solar thermal applications and is commercially available. Hitec salt has been used in several industries. Alternative salts (ternary mixtures) that can operate between 566°C and 121°C (1050°F and 250°F) have been identified, but additional research is necessary before large-scale demonstration is justified. Successful development of a TES system using these alternative salts could avoid the need for a combined molten salt and oil/rock TES system to cover the entire temperature range.

## System Descriptions

Design and performance characteristics were developed for the following six types of steam and/or electric power plants: 1) a boiler plant (boiler), as shown in Figure 21-2; 2) a simple-cycle electric power plant with steam cogeneration (SC/Cogen) as shown in Figure 21-3, 3) a simple cycle electric power plant with steam cogeneration and thermal

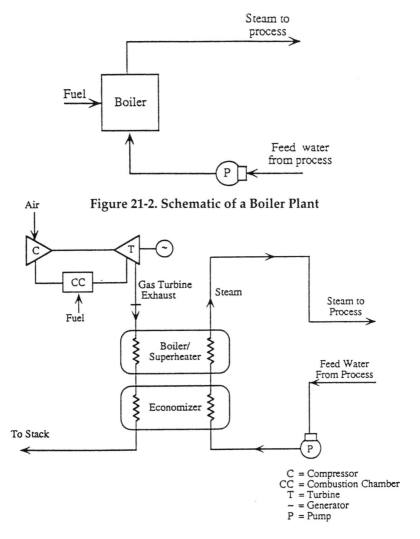

**Figure 21-2. Schematic of a Boiler Plant**

C = Compressor
CC = Combustion Chamber
T = Turbine
~ = Generator
P = Pump

**Figure 21-3. Schematic of a Simple-cycle Cogeneration Plant (SC/Cogen)**

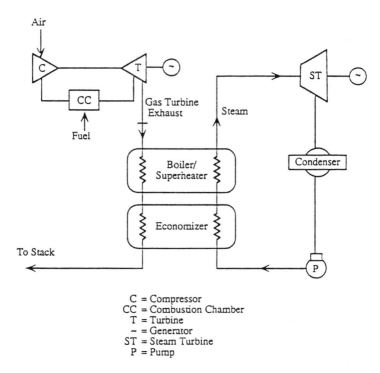

C = Compressor
CC = Combustion Chamber
T = Turbine
~ = Generator
ST = Steam Turbine
P = Pump

**Figure 21-4. Schematic of a Combined-cycle Electric Power Plant (CC)**

energy storage (SC/TES/Cogen) (Figure 21-1); 4) a combined-cycle electric power plant (CC) (Figure 21-4); 5) a combined-cycle electric power plant with steam cogeneration (CC/Cogen) (Figure 21-5); and 6) a combined-cycle electric power plant with steam cogeneration and thermal energy storage (CC/TES/Cogen) (Figure 21-6).

The first, second, and fourth plants were evaluated to provide a reference for comparing the cost of steam and electricity from the SC/TES/Cogen plant, while the first, fourth, and fifth plant concepts were evaluated to provide a reference for the CC/TES/Cogen plant. The boiler plant was evaluated to define the reference cost of producing steam, hence the value of steam produced by the cogeneration plants. Similarly, the CC plant was evaluated to define the reference cost of producing electricity, hence the value of electricity produced by the cogeneration plants. Many factors affect the value of products such as steam and electricity. This approach is consistent with defining value as equal to the marginal cost of production from the likely alternative

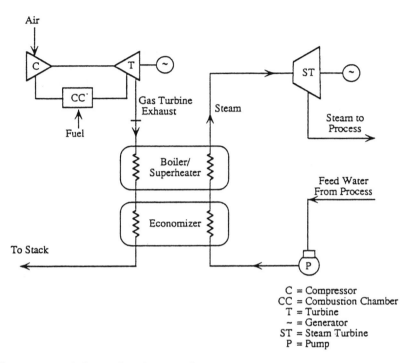

Figure 21-5. Schematic of a Combined-cycle Cogeneration Plant (CC/Cogen)

source. While a conventional boiler plant was an obvious reference technology for producing steam, many different options are available for producing electricity. A CC power plant provides a convenient benchmark for comparison with CC/Cogen and CC/TES/Cogen because the latter two plants are based on modifications to the basic CC power plant technology.

## PLANT ARRANGEMENTS

The base-case system is a conventional boiler plant (for setting the steam cost) and a combined-cycle power plant (for setting the electricity cost) against which a simple and a combined-cycle cogeneration system (with a gas-turbine prime mover and/or a steam turbine with a heat recovery steam generator (HRSG)) were compared.

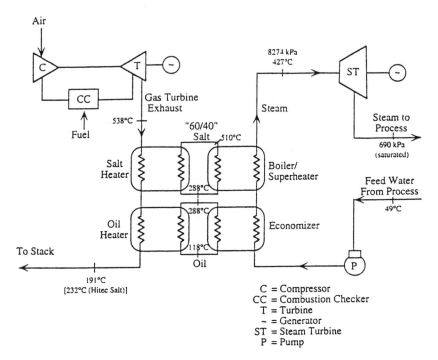

C = Compressor
CC = Combustion Checker
T = Turbine
~ = Generator
ST = Steam Turbine
P = Pump

**Figure 21-6. Schematic of a Combined-cycle Cogeneration Plant with Thermal Energy Storage (CC/TES/Cogen)**

### Conventional SC/Cogen Plant

The conventional simple-cycle cogeneration plant (Figure 21-3) consists of a gas turbine (GT) that is fired by a natural gas combustor. In addition to producing power through the generator, the turbine exhaust at 538°C (1000°F) is used in a heat recovery steam generator (HRSG) to produce the process steam load. The net efficiency of the gas turbine is assumed to be 31% (a heat rate of 11,000 Btu/kWh) for larger gas turbines (> 100 MWe rating) and 29.6% (a heat rate of 11,500 Btu/kWh) for smaller turbines (≤ 100 MWe rating).

### SC/Cogen Plant with TES for Peaking

An oil/rock or salt TES system interposed between the gas turbine and the steam generator in a conventional simple-cycle cogeneration plant can provide a cycling capability (Figure 21-1). Instead of generating steam directly, the heat from the gas turbine exhaust is used to heat the oil or molten salt, which is then stored in a tank until further use. The gas turbine is operated whenever peaking power capacity is needed by the utility. The TES serves to decouple the steam generator and the gas turbine from the rest of the plant, allowing process steam production for a steam turbine or other process thermal loads to continue for the entire day. In the case of the heat recovery salt heater, it may be possible to use direct-contact heat exchange between the exhaust gas and the salt. If feasible, this direct-contact heat exchange process would dramatically reduce the cost of the heat recovery salt heater and would improve the overall plant performance. There was no attempt made to optimize the sizes and costs of the components of the different systems evaluated in this study. Otherwise, more advantageous versions of each TES/cogeneration system or other system configurations with molten salt storage could have been designed and analyzed.

### Conventional CC/Cogen Plant

One of the primary goals of this study was to develop concept arrangements that minimize the impact of including TES on the design and the layout of a cogeneration plant. Therefore, there is substantial similarity between a conventional combined-cycle cogeneration plant (CC/Cogen) and the combined-cycle cogeneration with TES design (CC/TES/Cogen). The conventional combined-cycle cogeneration plant (Figure 21-5) consists of a prime mover (a gas turbine, in this case) that is fired by a natural gas combustor. The turbine exhaust at 538°C (1000°F) is used in a HRSG to produce steam loads of a power-producing steam turbine, which, in turn, exhausts to provide the process steam load. Therefore, the electrical capacity of this plant is greater than that of the simple-cycle cogeneration plant.

### CC/Cogen Plant with TES for Peaking

An oil/rock or salt TES system interposed between the gas turbine and the steam generator in a conventional combined-cycle cogeneration plant can provide a cycling capability (Figure 21-6) similar to the case of the simple-cycle cogeneration plant combined with thermal energy storage discussed earlier.

### System Steam Requirements

Process steam requirements are summarized in Table 21-1. Several systems were evaluated for meeting the same process steam load. In some cases, the systems are the same as one of six plant types described above, while others are combinations of two or more of the plant types. The gas turbine rating depends on the length of time during the day that the gas turbine is operated with intermediate- and/or peak-load electricity being sold to the utility. For example, having assumed the system will supply a constant 24-hr steam load, the rating is twice the base-load size if it were operating only for 12 hrs in a day. The waste heat recovery is in the form of heated oil or salt that is then stored in the oil/rock or salt storage tank to supply the 24-hr steam load. The additional system analyzed was the 8-hr operation of an oversized (threefold) gas turbine for selling peak power to the local utility. The alternative systems evaluated were 1) a boiler operating 24 hrs/day, 2) a SC/Cogen or a CC/Cogen plant operating 24 hrs/day, 3) a SC/Cogen or a CC/Cogen plant operating 8 hrs/day and selling peak-load power to the utility and a boiler operating for the remaining 16 hrs/day to supply the steam load, 4) a SC/Cogen or a CC/Cogen plant operating 12 hrs/day to sell intermediate-load power to the utility and a boiler operating 12 hrs/day, and 5) a SC/TES/Cogen plant using one of three TES systems discussed above or a CC/TES/Cogen plant using "60/40" salt and oil/rock storage or Hitec salt storage. The last alternative was evaluated with the gas turbine operating either 8 or 12 hrs/day and the steam turbine (in the CC/TES/Cogen plant) operating 24 hrs/day for both types of storage systems.

### Table 21-1. Process Steam Requirements

Flow Rate: 181,440 kg/hr (400,000 lb/hr); 24 hrs/day; 320 days/year

Supply Conditions: 690 kPa (100 psia); saturated steam with a quality of 0.973

Condensate Return
Conditions: 49°C (120°F) (saturated liquid)

---

**System Sizing**

Rudimentary design specifications were developed for each major system component to define the cost and performance basis. In general, equipment was sized to meet the steam requirements stated in Table 21-1. Key design and performance assumptions are presented in Table 21-2.

## Table 21-2. Design and Performance Assumptions

Natural Gas-Fired Systems

Steam Turbine Inlet Conditions:
- 181,440 kg/hr (400,000 lb/hr)
- 8274 kPa (1200 psia)
- 427 °C (800°F)

Steam Turbine Power Calculations
- turbine efficiency = 0.90
- generator efficiency = 0.98
- parasitic power = 0.02

Gas Turbine Heat Rate
- 11,500 Btu/kWh for GT sizes ≤ 100 MWe
- 11,000 Btu/kWh for GT sizes > 100 MWe

Gas Turbine Exhaust Temperature = 538°C (1000°F)

Overall Heat Transfer Coefficients
- 150 W/m$^2$ °C (26.6 Btu/hr ft$^2$ F) for HRSG and storage media heaters
- 846 W/m$^2$ °C (150 Btu/hr ft$^2$ F) for storage media steam generators

Storage Media Cycle Temperatures
- "60/40" salt: 288°C to 510°C (550 to 950°F)
- Oil/rock: 118°C to 288°C (250 to 550°F)
- Hitec salt: 218°C to 454°C (425 to 850°F)

*Steam Generator Sizing*

Steam generators include conventional gas turbine heat recovery steam generators and steam-generation equipment from thermal storage media. Process steam condensate and steam turbine inlet conditions define the economizer, boiler, and superheater heat duties and water/ steam inlet and exit temperatures. Sizing of these units depended on both the gas turbine exhaust temperature and the ultimate reject temperature after heat recovery. In general, lowering the reject temperature increases the waste heat recovery fraction and reduces the size of the gas turbine required, but results in larger, more costly heat exchangers. The minimum reject temperature is limited by the boiler pinch point. A reasonable ultimate reject temperature was selected from several investigated, but a formal optimization was not conducted.

The first step for sizing storage media heated steam generators was to select the media operating temperature range from within the upper and lower temperature limits. In general, the temperature range should be as large as possible to minimize storage costs. A higher upper temperature will reduce steam generator costs but increase media heater costs. Boiler pinch point limitations must also be considered in setting the lower media temperature. Thus, the operating temperature range affects all TES charging and discharging equipment, as well as the TES unit. Again, the design approach was to select reasonable, but not necessarily optimal temperature ranges; the specific temperature range for each media type was shown in Table 21-2. Once the media temperature range was established, the design procedure was the same as that described for the HRSG.

*Storage Sizing*

Thermal storage capacity (MWht) is independent of the media type because the same total energy must be transferred in the steam generator, and the storage efficiency is essentially the same. Thermal losses for large (500 to 2,500 MWht) storage systems such as those required for the systems evaluated in this study are less than 1% (Williams et al. 1987). An overall efficiency of 97%, which allows for losses in piping and the thermal equivalent of pumping parasitics, was presumed. The required storage capacity is directly proportional to the steam generation energy and the number of hours the steam generator is operated from storage (or 24 minus the number of hours the gas turbine operates).

*Storage Media Heater Sizing*

Media heater thermal duties and media temperatures were established as part of the media heated steam generator sizing process. Design considerations and procedures were similar to those described above for the HRSG sizing. In general, the minimum gas turbine exhaust reject temperature is limited by the minimum media temperature. For the nitrate salt and oil/rock storage system, consideration must also be given to the pinch point at the low-temperature end of the salt heater and high-temperature end of the oil heater.

*Gas Turbine Sizing*

The required gas turbine generating capacity depends on its heat rate and the exhaust temperature (after thermal recovery in the HRSG or media heater). Heat rates are normally quoted in Btu/kWh. For example, gas turbines with a generating capacity greater than 100 MWe were assumed to have a heat rate of 11,000 Btu/kWh in this study. If inputs and outputs are both expressed in kWh, the equivalent heat rate is 3.223, i.e., 3.223 kWh of fuel energy are converted into 1.000 kWh of electricity and 2.223 kWh of exhaust heat. In this study, all exhaust heat was assumed to leave the turbine in the form of the exhaust gas at a temperature of 538 °C (1000 °F). Thus, the ratio of electric energy to exhaust energy is 1/2.223 or more generally, 1/(HR-1), where "HR" is the heat rate in kWh of fuel energy per kWh of electricity. Recoverable energy in the exhaust is measured relative to 25°C (77°F), the reference temperature for measuring the energy input from the gas fuel. Therefore, the heat recovery fraction becomes (1000-TR)/(1000-77), where "TR" equals the ultimate reject temperature after heat recovery. The waste heat recovery fraction can be combined with the electric/exhaust energy ratio to produce Equation (1) defining the relationship between gas turbine generating capacity and the waste heat recovery rate. The maximum capacity of an individual gas turbine unit was limited to 150 MWe, resulting in either two or three parallel gas turbine and heat recovery trains.

$$kWe = kWt * 923/[(HR - 1) * (1000 - TR)] \tag{1}$$

*Equipment Sizes*

Gas turbine, steam turbine, media heater, storage, and steam generator equipment sizes are summarized in Table 21-3.

# ECONOMIC EVALUATION

The economic evaluation was conducted by calculating and comparing the levelized cost of steam produced by the alternative concepts being considered. Levelized cost analysis combines initial cost, annually recurring cost, and system performance characteristics with financial pa-

**Table 21-3. System Equipment Sizes**

| Systems | Size/Rating |
|---|---|
| Boiler | 181,440kg/hr (400,000 lb/hr) steam. |
| **Simple-Cycle Cogeneration (SC/Cogen)** | |
| Gas Turbine, MWe | 89 |
| Heat Recovery Steam | |
| Generator (HRSG), $m^2$ | 4627 |
| **Combined-Cycle (CC)** | |
| Gas Turbine, MWe | 94 |
| HRSG, $m^2$ | |
| Economizer | 5846 |
| Boiler | 6143 |
| Superheater | 1041 |
| Steam Turbine, MWe | 50.0 (condensing) |
| **Combined-Cycle Cogeneration (CC/Cogen)** | |
| Gas Turbine, MWe | 94 |
| HRSG, $m^2$ | |
| Economizer | 5846 |
| Boiler | 6143 |
| Superheater | 1041 |
| Steam Turbine, MWe | 50.0 (condensing) |
| | Size/Rating |

| Simple-Cycle Co-generation with TES (SC/TES/Cogen) | Oil/Rock | | "60/40" Salt | | Hitec Salt | |
|---|---|---|---|---|---|---|
| | 12 hr GT Operation | 8 hr GT Operation | 12 hr GT Operation | 8 hr GT Operation | 12 hr GT Operation | 8 hr GT Operation |
| Gas Turbine, MWe | 2 × 97 | 3 × 97 | 2 × 138 | 3 × 138 | 2 × 97 | 3 × 97 |
| Media Heater, m² | 13012 | 20772 | 102048 | 153066 | 36547 | 54818 |
| Storage, MWht | 1874 | 2500 | 1953 | 2604 | 1562 | 2083 |
| Steam Generator, m² | 3234 | 3234 | 649 | 649 | 1103 | 1103 |

| Simple-Cycle Co-generation with TES (SC/TES/Cogen) | Oil/Rock and "60/40" Salt | | Hitec Salt | |
|---|---|---|---|---|
| | 12 hr GT Operation | 8 hr GT Operation | 12 hr GT Operation | 8 hr GT Operation |
| Gas Turbine, MWe | 2 × 103.5 | 3 × 103.3 | 2 × 117.5 | 3 × 117.3 |
| Steam Turbine, MWe | 24.4 | 24.4 | 24.4 | 24.4 |
| Media Heater, m² | | | | |
| Oil/Rock | 2 × 9201 | 3 × 9201 | | |
| "60/40" salt | 2 × 32528 | 3 × 32528 | | |
| Hitec salt | | | 2 × 26580 | 3 × 26580 |
| Storage, MWht | | | | |
| Oil/rock | 597 | 795 | | |
| "60/40" salt | 1233 | 1644 | | |
| Hitec salt | | | 1830 | 2440 |
| Steam Generator, m² | | | | |
| Economizer | 2379 | 2379 | 1236 | 1236 |
| Boiler | 1840 | 1840 | 1989 | 1989 |
| Superheater | 223 | 223 | 428 | 428 |

rameters to produce a single figure-of-merit (the levelized cost) that is economically correct and can be used to compare the projected steam costs of alternative boiler and cogeneration plant concepts. The specific methodology used was that defined in Brown et al. (1987).

Specific financial assumptions used to calculate the levelized steam cost are listed in Table 21-4. These assumptions are intended to be representative of industrial ownership. Brown et al. (1987) was the reference for the discount rate, general inflation rate, property tax and insurance rate, and combined state and federal income tax rate. The economic life was set at 30 years based on standards prescribed by the Electric Power Research Institute (EPRI) (1989) for facilities similar to the boiler and cogeneration plants considered in the current study. The corresponding depreciable life is 20 years (Van Knapp et al. 1989). The first year of operation was set at 1995 because the storage systems considered in the current study are mature and could be implemented immediately. The price year was set to mid-1990 for convenience. The system construction period, set at 2 years, was also based on data presented in EPRI (1989) for similar systems. Capital and nonfuel operation and maintenance (O&M) costs were assumed to escalate at the same rate as general inflation. Natural gas was assumed to escalate at 3.8% in excess of general inflation (i.e., at 7%/year overall) based on fuel price projections prepared by the Energy Information Administration (1991).

## Table 21-4. Financial Assumptions

| Description | Assumption |
|---|---|
| System economic life | 30 years |
| System depreciable life | 20 years |
| Nominal discount rate | 9.3%/year |
| General inflation rate | 3.1 %/year |
| Capital inflation rate | 3.1 %/year |
| O&M inflation rate | 3.1 %/year |
| Natural gas inflation rate | 7.0%/year |
| Combined state and federal income tax rate | 39.1% |
| Property tax and insurance rate | 2.0% |
| System construction period | 2 years |
| Price year | 1990 |
| First year of system operation | 1995 |

In general, a levelized cost analysis determines the revenue required to exactly cover all costs associated with owning and operating a facility, including return on investment. Typically, the required revenue is expressed per unit of production, e.g., $/kWh or $/klb steam. For cogeneration systems, there are two revenue producing products, electricity and steam. Increasing the revenue associated with electricity decreases the revenue required from steam and vice-versa. For the simple-cycle cogeneration analysis, the reference cost of steam was established on the basis of the boiler plant, and for the combined-cycle analysis, either the electric or steam revenue rate was assumed for each cogeneration case and the levelized cost analysis solved for the required revenue rate for the other product. The value of cogenerated steam and electricity was established based on the cost of steam from a stand-alone boiler plant and the cost of electricity from a combined-cycle plant. Capital cost and operation and maintenance cost estimates were based on models developed for the steam and cogeneration plant components (Somasundaram et al. 1992).

# RESULTS AND DISCUSSION

The break-even electric rates for the simple-cycle cogeneration analysis at which the levelized steam cost is the same as that of a boiler plant ($9.03/klb) are given in Table 21-5. The break-even rate for the conventional cogeneration system is $0.035/kWh, while that for the cogeneration system with TES varies depending on the storage medium and the power production schedule. The corresponding rate for a gas turbine plant is given for comparison purposes. It can be seen that the oil/rock TES system can provide on-peak power at a cost of $0.045/kWh to $0.050/kWh, which is considerably less expensive than the simple gas turbine case (or the CC plant as shown in Table 21-6). The molten salt cases are less attractive for the assumed process load conditions. The Hitec salt can provide peak power at a slightly less expensive rate than the "60/40" salt, primarily because of the wider temperature range of the storage medium. In general, lower-temperature storage reduces the size and cost of the storage media heater, while higher-temperature storage reduces the size and cost of the media-heated steam generator. Poor heat transfer in the media heater (on the exhaust gas side) puts a premium on the lower approach temperatures required of high-temperature storage

**Table 21-5. Breakeven Electric Rates for Boiler Steam Costs (Levelized)**

| System Configuration | Breakeven Electric Rates ($/kWh) | |
|---|---|---|
| Simple-Cycle Gas Turbine (SC) | 0.08 | |
| Simple-Cycle Cogeneration (SC/Cogen) (24-hr GT operation) | 0.035 | |
| Simple-Cycle Cogeneration with TES (SC/TES/Cogen) | | |
| | 12-hr GT operation | 8-hr GT operation |
| Oil/Rock | 0.045 | 0.050 |
| "60/40" Salt | 0.079 | 0.095 |
| Hitec Salt | 0.059 | 0.070 |

**Table 21-6. Reference Steam and Electricity Costs/Values**

| | Daily Operating Period | | |
|---|---|---|---|
| | 8 hrs | 12 hrs | 24 hrs |
| Boiler LEC, $/klb | 11.23 | 10.01 | 8.71 |
| CC Plant LEC, $/kWh | 0.072 | 0.064 | 0.055 |

systems. Thus, the oil/rock system has a heat exchanger sizing and cost advantage over the two salt systems. The oil/rock system is also the least expensive (on a $/MWht basis) when each storage system is allowed to cycle through its maximum temperature range. It should also be noted that the systems evaluated have not been optimized; more advantageous versions of each TES/cogeneration system could be identified by considering other combinations of storage media temperature range and heat exchanger approach temperatures. Varying these design factors trades off heat exchanger and storage system costs. Also, future research and development efforts focused on the salt storage media may further reduce the costs of such storage media and make them more attractive for wider range of temperature conditions.

For the case of combined-cycle cogeneration, the levelized costs of steam production from a boiler plant, and that of electricity production from a CC plant were calculated to establish the reference cost/value of these two commodities when produced by a cogeneration plant (Somasundaram et al. 1993). The levelized energy costs (LECs) for these two systems are shown in Table 21-6 for different daily operating periods. As would be expected, the LECs decline with increased daily operating hours as the fixed capital and O&M costs are spread over a larger annual energy output.

The LECs from a CC/Cogen plant were also calculated to establish a reference for measuring the impact of adding TES. Again, for any multiple energy product operation, the value of all but one energy product must be fixed to solve for the LEC of the remaining product. Thus, the LECs of steam and electricity from a CC/Cogen plant were calculated by alternately fixing the value of steam or electricity at the levels indicated in Table 21-6. Table 21-7 presents steam LECs for alternative systems producing identical rates of steam flow 24 hrs/day, with electric power being produced at different schedules and amounts. The results indicate that a CC/Cogen plant operating 24 hrs/day would produce steam at the lowest possible cost. In addition, a CC/Cogen/boiler hybrid system would produce steam at a lower average cost than a stand-alone boiler, as long as the CC/Cogen part of the system is operated for at least 8 hrs/day. (Note: At some daily operating period of less than 8 hrs, a stand-alone boiler would be preferred, but this break point was not determined.)

**Table 21-7. Levelized Cost of Baseload Steam**

| System Description | Levelized Energy Cost, $/klb |
|---|---|
| 24 hr boiler | 8.71 |
| 24 hr CC/Cogen | 2.90 |
| 12 hr CC/Cogen and 12 hr boiler | 6.73 |
| 8 hr CC/Cogen and 16 hr boiler | 7.62 |
| 12 hr CC/Hitec Salt TES/Cogen | 8.46 |
| 8 hr CC/Hitec Salt TES/Cogen | 10.77 |

Table 21-8 presents electricity LECs for alternative systems producing power 8 or 12 hrs/day at different rates with steam being produced at identical rates 24 hrs/day. Finally, Table 21-9 shows the marginal cost of electric power provided by the CC/Hitec Salt TES/Cogen system relative to the reference CC/Cogen system. The results shown in these two tables further emphasize that the CC/Cogen system is the preferred option to the CC/TES/Cogen systems. In fact, the marginal cost of power from the CC/Hitec Salt TES/Cogen system is significantly higher than the cost of power produced by a CC plant alone.

**Table 21-8. Levelized Cost of Peaking Power**

| System Description | Levelized Energy Cost, $/kWh | |
|---|---|---|
| | 8-hr peak | 12-hr peak |
| CC | 0.072 | 0.064 |
| CC/Cogen | 0.061 | 0.050 |
| CC/Hitec Salt TES/Cogen | 0.079 | 0.063 |

**Table 21-9. Marginal Cost of Peaking Power**

| System Description | Levelized Energy Cost, $/kWh | |
|---|---|---|
| | 8-hr peak | 12-hr peak |
| CC/Hitec Salt TES/Cogen | 0.105 | 0.099 |

# ENVIRONMENTAL CONSIDERATIONS

Atmospheric emissions for cogeneration plants result from the combustion of fuel in the gas turbine. The principal concern for natural-gas-fired turbines is nitrogen oxides (NOx), although other components (carbon monoxide (CO) and total solid particulates (TSP)) are also emitted, albeit in relatively minor amounts. Actual emissions will depend on the design of the turbine and the operating conditions. The emissions reported here for cogeneration systems are based on data presented in Esposito (1989) for a General Electric Model PG-7 111EA turbine used in a combined-cycle application. Reference operating conditions and emission rates are presented in Table 21-10.

Emissions data for six different cogeneration systems considered here are presented in Table 21-11. Fuel consumption, steam production, and electricity production numbers for each of these systems is presented in Table 21-12 to help explain the emissions results. Total emis-

**Table 21-10. Gas Turbine Emission Assumptions**

Turbine Type: General Electric Model PG-7111EA
Ambient Temperature: 15°C (59°F)
Ambient Humidity: 60% relative humidity
Site Elevation: sea level
$NO_x$ emission concentration: 42 ppmv
$NO_x$ control method: steam injection
Fuel: natural gas
$NO_x$ emission rate = 0.535 kg/MMBtu
CO emission rate = 0.036 kg/MMBtu
TSP emission rate = 0.018 kg/MMBtu

sions are directly proportional to annual fuel consumption, which varies depending on the design of each cogeneration system. The combined-cycle (CC) systems consume more fuel, but produce more power than the simple-cycle (SC) systems. Similarly, turbine cycles with TES systems consume more fuel, but produce more power than non-TES systems with the same cycle. Emissions per unit of steam production (kg/klb) follow exactly the same pattern as total emissions because steam production is the same for all systems. When emissions are presented per unit of electricity production (kg/MWh), however, the rankings change. The CC systems have lower emissions than SC systems, and TES systems have lower emissions than non-TES systems with the same cycle. The rankings can be explained by the lower heat rates of CC systems compared to SC systems, and the lower heat rates of larger turbines used for the TES systems compared to the nonTES systems. This latter advantage is offset by relatively poor waste heat recovery for the CC/TES application, however. Therefore, emissions per MWh for TES and non-TES CC systems are about the same.

# SUMMARY AND CONCLUSIONS

Thermal energy storage can help cogeneration meet the challenges of the 1990s by increasing the flexibility and performance of cogeneration facilities. Thermal energy storage also allows a cogeneration facility to provide dispatchable electric power while providing a constant thermal

## Table 21-11. Cogeneration System Emissions

| System Description | NO$_x$ Emissions | CO Emissions | TSP Emissions | Units |
|---|---|---|---|---|
| 24 hr CC/Cogen | 4450 | 301 | 150 | Mg/yr |
|  | 1.449 | 0.098 | 0.049 | kg/klb |
|  | 4.886 | 0.330 | 0.165 | kg/MWh |
| 8 hr CC/Hitec Salt | 5302 | 358 | 179 | Mg/yr |
| TES/Cogen | 1.726 | 0.117 | 0.058 | kg/klb |
|  | 4.871 | 0.329 | 0.165 | kg/MWh |
| 12 hr CC/Hitec Salt | 5309 | 359 | 179 | Mg/yr |
| TES/Cogen | 1.728 | 0.117 | 0.058 | kg/klb |
|  | 4.872 | 0.329 | 0.165 | kg/MWh |
| 24 hr SC/Cogen | 4181 | 282 | 141 | Mg/yr |
|  | 1.361 | 0.092 | 0.046 | kg/klb |
|  | 6.151 | 0.416 | 0.208 | kg/MWh |
| 8 hr SC/(Oil/Rock) | 4391 | 297 | 148 | Mg/yr |
| TES/Cogen | 1.429 | 0.097 | 0.048 | kg/klb |
|  | 5.884 | 0.398 | 0.199 | kg/MWh |
| 12 hr SC/(Oil/Rock) | 4390 | 297 | 148 | Mg/yr |
| TES/Cogen | 1.429 | 0.097 | 0.048 | kg/klb |
|  | 5.884 | 0.398 | 0.199 | kg/MWh |

load. The first of two studies reported here focused on the relative performance and economic benefits of incorporating a diurnal TES system with a simple-cycle gas turbine cogeneration system to produce dispatchable power during peak- and/or intermediate-demand periods. The results showed that the conventional cogeneration system and the cogeneration plant combined with oil/rock TES produced steam at a lower cost than a conventional boiler plant operation as long as the electricity sale price remained above $ 0.06/kWh. The break-even electricity price (at which the steam costs are the same for the different plant con-

**Table 21-12. Cogeneration System Energy Inputs and Outputs**

| System Description | Natural Gas MMBtu/year | Steam klb/year | Electricity MWh/year |
|---|---|---|---|
| 24 hr CC/Cogen | 8,319,744 | 3,072,000 | 910,863 |
| 8 hr CC/Hitec Salt TES/Cogen | 9,912,320 | 3,072,000 | 1,088,527 |
| 12 hr CC/Hitec Salt TES/Cogen | 9,926,400 | 3,072,000 | 1,089,807 |
| 24 hr SC/Cogen | 7,816,320 | 3,072,000 | 679,680 |
| 8 hr SC/(Oil/Rock) TES/Cogen | 8,208,640 | 3,072,000 | 746,240 |
| 12 hr SC/(Oil/Rock) TES/Cogen | 8,207,232 | 3,072,000 | 746,112 |

figurations) is $0.035/kWh for the conventional cogeneration case, and $0.045 to $0.05/kWh for the cogeneration system combined oil/rock TES. This represents nearly a 40% reduction in the cost of peak power when compared to $0.08/kWh for a gas turbine plant; and a 30% reduction compared to a peak power cost of approximately $0.07/kWh for a combined-cycle plant. The oil/rock storage system for TES was found to be the most attractive option for the assumed thermal load quality. A higher quality of the assumed thermal load (e.g., at higher pressures and temperatures) was also explored (saturated steam at 3450 kPa or 500 psia) and the oil/rock TES still remained as the attractive option. The molten salt systems may become more attractive in the future if the media costs are found to be lower than the assumed levels.

The second study evaluated the cost of power produced by a combined-cycle electric power plant (CC), a combined-cycle cogeneration plant (CC/Cogen) and a combined-cycle cogeneration plant integrated with thermal energy storage (CC/TES/Cogen). The two cogeneration systems were designed to serve the same process steam load. The value of producing electricity was set at the levelized cost for the CC plant, while that of the process steam was for a conventional stand-alone boiler. The results indicate relatively poor economic prospects for integrating TES with a combined-cycle cogeneration power plant. The biggest part of the problem can be attributed to the extremely close approach temperatures at the storage media heaters, which makes them large and expensive. Two potentially mediating factors should, however, be considered prior to formulating any final conclusions or hastily eliminating this system combination. First, the designs developed here were reasonably practical and would work, but were not optimized from a size or cost perspective. For example, increasing the approach temperatures for the storage media heaters might lower the media heater costs more than it would raise media storage costs. Quite simply, design optimization would improve the economics, perhaps enough to make the CC/TES/ Cogen an attractive option for incremental peak power production. Second, the media heaters were based on conventional "shell-and-tube" type heat transfer equipment. Direct-contact of gas turbine exhaust with the storage media would permit a much closer approach temperature, while reducing the cost of the media heater by as much as a factor of five. This magnitude of cost reduction for the high-temperature media heater could result in economically more attractive CC/TES/Cogen applications. Therefore, further analysis of this system combination, and especially using direct-contact heat transfer equipment for media heating, is strongly recommended.

## References

Brown, D.R., J.A. Dirks, M.K. Drost, G.E. Spanner, and T.A. Williams. 1987. *An Assessment Methodology for Thermal Energy Storage Evaluation*. PNL-6372, Pacific Northwest Laboratory, Richland, Washington.

Drost, M.K., Z.I. Antoniak, D.R. Brown, and K. Sathyanarayana, 1989. *Thermal Energy Storage for Power Generation*. PNL-7107, Pacific Northwest Laboratory, Richland, Washington.

Drost, M.K., Z.I. Antoniak, D.R. Brown, and S. Somasundaram. 1990. *Thermal Energy Storage for Integrated Gasification Combined-Cycle Power Plants*. PNL-7403, Pacific Northwest Laboratory, Richland, Washington.

Electric Power Research Institute. 1989. *Technical Assessment Guide. Electric Supply 1989*. EPRI P-6587-L. Electric Power Research Institute,

Palo Alto, California.

Energy Information Administration. 1991. *Annual Energy Outlook. With Projections to 2010*. DOE/EIA-0383(91). Energy Information Administration, Washington, D.C.

Esposito, N.T. 1989. *A Comparison of Steam-Injected Gas Turbine and Combined-Cycle Power Plants: Technology Assessment*. EPRI GS-6415. Electric Power Research Institute. Palo Alto, California.

Somasundaram, S., D.R. Brown, and M.K. Drost. 1992. *Evaluation of Diurnal Thermal Energy Storage Combined with Cogeneration Systems*. PNL-8298, Pacific Northwest Laboratory, Richland, Washington.

Somasundaram, S., D.R. Brown, and M.K. Drost. 1993. *Evaluation of Diurnal Thermal Energy Storage Combined with Cogeneration Systems - Phase II*. PNL-8717, Pacific Northwest Laboratory, Richland, Washington.

Van Knapp, D.P., J.D. Perovich, L.E. Griffith, Jr., S.A. Bock, and M.S. Waldman, eds. 1989. *American Jurisprudence. Second Edition*. The Lawyers Co-operative Publishing Company, Rochester, New York, and Bancroft-Whitney Company, San Francisco, California.

Williams, T.A., J.A. Dirks, D.R. Brown, M.K. Drost, Z.A. Antoniak, and B.A. Ross. 1987. *Characterization of Solar Thermal Concepts for Electricity Generation*. PNL-6128, Vols. 1 and 2, Pacific Northwest Laboratory, Richland, Washington.

*Chapter 22*

# Encouraging Efficient Cogeneration: The State Role

*Chapter 22*

# Encouraging Efficient Cogeneration: The State Role

*Barney L. Capehart, Ph.D.*
*Lynne C. Capehart, J.D.*
*University of Florida*
*William D. Orthwein*

G rowing concern over such issues as global climate change, meeting the requirements of the Clean Air Act Amendments of 1990, and economic and security problems as a result of purchasing imported fuels should provide a national motivation for developing highly efficient cogeneration.

However, as the authors have shown previously, many of the cogeneration facilities currently being constructed have very low efficiency levels. This means that the actual fuel savings from cogeneration and the affiliated reduction in sulfur dioxide and other air emissions per kilowatt-hour generated are substantially lower than the maximum levels that should be achieved [*Public Utilities Fortnightly*, March 15, 1990].

Under current Federal Energy Regulatory Commission (FERC) rules, a cogeneration facility is considered a qualifying facility (QF) (and thus eligible for incentive payments under the Public Utilities Regulatory Policy Act) if it uses at least 5% of its thermal energy for some purpose besides power generation. Additional requirements exist for cogenerators using natural gas or fuel oil.

This standard for Utilized Thermal Energy Production (UTEP) is not a very demanding standard since UTEPs up to 85 percent are common for industrial cogeneration. However, a large percentage of today's cogenerators are limiting themselves to the bottom end of the UTEP standard [*Public Utilities Fortnightly*, April 1, 1991].

The easily met FERC standard has thus had the unfortunate result of shifting the primary purpose of cogeneration. Before PURPA, most cogenerators supplied electric power as a by-product of their industrial steam process; after PURPA, most cogenerators supply industrial steam as a by-product of their power generation process.

Thus, under PURPA, many cogenerators are really power plants that supply only enough thermal energy to qualify for avoided cost payments under the FERC rules. To guarantee that cogenerators produce acceptable energy and environmental benefits in return for the financial and regulatory incentives they receive, their minimum efficiency levels must be increased. Two major approaches to increasing cogeneration efficiency provide potential remedies—one at the federal level, and one at the state level.

The current regulatory environment for the generation of electrical power is such that many more independent power producers and QFs will be built in the near future. As deregulation of power production proceeds, emphasis will be on reducing the cost of producing power rather than insuring that higher overall thermal efficiencies of these facilities are achieved.

Just as many electric utilities are now cutting their efforts and expenditures in the DSM area, there will be similar reductions in efforts and expenditures to incorporate the costs of environmental externalities into the economic decision to build or to buy power from high efficiency cogeneration facilities. However, the opportunity to promote high efficiency cogeneration is still readily available at both the federal and state levels if they choose to do so.

# WHO SHOULD PROMOTE
# COGENERATION EFFICIENCY?

Although cogeneration efficiency was once primarily the concern of industry, corrective government action is now necessary to undo prior actions which have artificially stimulated low-efficiency cogeneration. The logical body to act is the federal government which could reverse the current trend by increasing the present 5% FERC standard for Qualifying Facilities through a rule change or a legislative directive. The authors have already discussed this option in an earlier article, and proposed a standard of 50% as a more appropriate level [Public Utilities

*Fortnightly*, April 1, 1991].

However, there is little movement at the federal level to correct these cogeneration efficiency problems, since the present congressional emphasis on minimizing government regulation leaves any corrective actions up to the states. The 1992 Energy Policy Act which implemented the National Energy Strategy dealt with many issues of electricity regulation, but did not include a revision to the FERC standard for cogenerators.

In the absence of federal action, states themselves could stimulate higher efficiency cogeneration through state legislative and administrative actions. Several states have already examined this issue and some have taken steps to address it directly.

Many other states have developed Integrated Resource Planning (IRP) or Least Cost Utility Planning (LCUP) processes; these processes provide an ideal forum for examining cogeneration efficiency. This chapter analyzes examples of representative state actions addressing cogeneration efficiency and suggests ways that states can best achieve the desired result.

Implementation of the provisions of the Clean Air Act Amendments of 1990 (CAAA) gives an added emphasis for states to take action to require increased efficiency cogenerators. Title IV of the CAAA sets requirements for sulfur dioxide ($SO_2$) reductions for electric power plants, and dictates a permanent ceiling of $SO_2$ emissions at 8.95 million tons per year nationwide.

This limit is also characterized as 8.95 million allowances—where one allowance is the authorization to emit one ton of $SO_2$ annually. To the extent that higher efficiency cogeneration reduces $SO_2$ emissions per kWh generated, utilities which purchase electricity from efficient cogenerators will not need to purchase as many allowances to cover their total $SO_2$ emissions. Thus, total electric generation costs can be held down.

# ENCOURAGING EFFICIENT COGENERATION AT THE STATE LEVEL

States can approach the problem of promoting efficient cogeneration either directly or indirectly. The most direct action for a state is to set its own efficiency standard for cogenerators considerably higher than the FERC standard. Promoting efficient cogeneration through incentives or disincentives is an indirect approach.

Many states have begun to develop Integrated Resource Planning or Least Cost Planning and competitive bidding processes; these processes provide ideal forums for indirectly promoting cogeneration efficiency.

This chapter examines the methods used in those states which directly encourage efficient cogeneration. It also looks at examples of other policies and rules which are related to efficiency of cogeneration and which would allow a state to achieve the goal of higher efficiency cogeneration with modification of existing rules. It also looks at private incentives for efficient cogeneration. Finally, the authors suggest model rules which states can adopt to promote cogeneration efficiency.

# DIRECT STATE ACTION: COGENERATION EFFICIENCY STANDARD

The most direct method for promoting efficient cogeneration is to set a standard for cogeneration efficiency that is substantially higher than the present FERC standard. This method was used by the state of Connecticut which is apparently the only state to have an actual cogeneration efficiency requirement that exceeds the FERC standard.

The Connecticut legislature set the higher standard in a statute defining a private power production facility. With respect to cogeneration facilities, the statute defines a "private power production facility" as "a facility which generates electricity in the state solely through the use of cogeneration technology, provided the average useful thermal energy output (UTEP) of the facility is at least 20% of the total energy output of the facility" [Connecticut General Statutes, Chapter 283, Section 16-243b(1)].

In Connecticut, unless a power producer is a private power producer, it is subject to regulation as a utility. (See Connecticut General Statutes, Chapter 283, Sections 16-243b(2) and 16-243a(b)(1); Chapter 277, Section 16-1a(8)). Thus, any cogenerator who wants to avoid being classed as a regulated utility in Connecticut must meet the 20% standard and be substantially more efficient than those permitted under the FERC rule.

However, this is still not a very rigorous requirement. As shown in our earlier report, the majority of both utility-owned and nonutility-owned cogeneration facilities have UTEPs in the range of 5 to 25 percent.

From our research, Connecticut appears to be the only state that has specifically recognized the problem with low efficiency cogeneration, and has directly adopted a requirement that cogenerators in the state meet stricter conditions on the UTEP Operating Standard then that included in PURPA. The Connecticut Statutes on cogeneration set up definitions and terminology, as well as substantive rights, which are significantly different from PURPA and FERC rules.

For all practical purposes, a cogenerator in Connecticut must meet state requirements to be classed as a Private Power Producer, as well as meeting the requirements of FERC rules to be a QF [Paul R. McCary, Cogeneration in Connecticut, *Connecticut Bar Journal*, Volume 59, 1985]. From the Connecticut experience, this direct approach should be transferable to other states, if they choose to adopt it.

# INDIRECT STATE ACTION: INCENTIVES FOR EFFICIENT COGENERATION

Cogenerators may try to argue that the state cannot set a standard different from the federal standard. Although PURPA does not appear to preclude states from setting higher standards, states can avoid this issue by providing incentives for more efficient cogeneration rather than mandating it.

A variety of incentives are available to the states. These incentives can be provided indirectly through the purchasing utilities or aimed directly at the cogenerator. Examples of utility-based incentives include rules which stimulate the purchase of efficiently cogenerated electricity through higher returns on investment, competitive bidding standards, or standard contracts.

Utilities could also be encouraged to purchase from the most environmentally benign cogenerator and required to include environmental externalities in their integrated resource planning. Direct incentives could be given to cogenerators through pollution taxes assessed on pollutants per kilowatt-hour.

* **Return on Investment**—To stimulate efficient cogeneration, public utility regulators could give utilities a higher rate of return for their

investment in efficient cogeneration facilities. The return on investment (ROI) could be scaled proportionate to the efficiency of the cogenerator or the higher ROI could simply apply to cogenerators whose UTEP met a high state efficiency standard.

• **Standard Contracts**—Standard contracts give both utilities and cogenerators some assurance of what contract terms the utility regulators find acceptable without additional negotiation. Thus, standard contracts provide another way to promote efficiency. Commissions could place provisions in the standard contracts which would quickly approve purchases from high efficiency cogenerators.

• **Buyback Rates**—A utility is supposed to pay a buyback rate that is equivalent to avoided cost for power produced by a cogenerator (QF). FERC defines avoided cost "as the incremental cost to an electric utility of electric energy or capacity, or both, which, but for the purchase, such utility would generate itself or purchase from another source." Some states have used the avoided cost requirement to establish incentives for utilities to purchase power generated from sources considered to be socially desirable.

• **Competitive Bidding Rules**—Competitive bidding procedures provide another option for states to encourage higher efficiency. Under state bidding procedures, utilities use a competitive system to evaluate and select QFs from which they will purchase energy or capacity on a long-term basis. The competitive bidding approach is considered a substitute for administrative determinations of avoided costs and represents a market-driven basis for setting payments to cogenerators.

The competitive bidding process can be used to encourage efficient cogeneration by setting low air emissions as a selection criteria. More efficient cogenerators would be able to meet lower air pollution emissions without costly emissions control equipment. Thus, cogenerators who had lower emissions because of higher efficiency would be more likely to be selected as successful bidders.

• **Integrated Resource Planning**—IRP requires utilities to combine their assessment of future energy supply and demand, taking into account energy efficiency and load-management programs, environmental and social factors, and the uncertainties and risks posed by different resource choices.

* **Environmental Incentives**—Environmental considerations provide additional options for the states. Either a direct emission tax, or a tax credit for reduced emissions, to account for environmental externalities could result in higher efficiency cogeneration. Incentives to utilities for use of renewable energy and noncombustion credits could include incentives for purchase from highly efficient cogenerators. Finally, utility regulators should require utilities to incorporate the efficiency of cogeneration as a consideration of their integrated resource plans.

A number of states have either added or considered environmental degradation as a factor for their bidding criteria. In addition, a number of utilities have implemented bidding systems in which environmental impacts of a bidder's project are evaluated explicitly in a weighting or point scheme.

A recent FERC order related to issues brought by Southern California Edison Company and San Diego Gas and Electric Company affirmed that states could only set avoided cost rates to include costs which actually would be incurred by utilities [71 FERC 61,269, June 2, 1995, Docket Number EL95-16-001 and Docket Number EL95-19-001]. In addition, FERC also clarified its views on the scope of state authority—both within and outside PURPA—to make resource planning decisions and to encourage renewable or alternative sources of generation.

In particular, FERC noted that resource planning and resource decisions remain the prerogative of state commissions and that states may wish to diversify their generation mix to meet environmental goals in a variety of ways. States are not barred from accounting for environmental costs of fuel sources included in an all-source determination of avoided costs.

Several specific ways that states could accomplish the goal of including environmental externalities were stated by FERC. One was a tax or other charge on all generation produced by a particular fuel; and another was to subsidize certain types of generation such as wind and other renewables through a tax credit. Thus this FERC order clearly stated that states had the right to include environmental factors in setting avoided cost rates; and even went out of its way to say how that could be done.

# EXAMPLES OF INDIRECT STATE ACTIONS

The following examples of incentives used or considered by selected states shows the wide range of initiatives available to states if they choose to use them. These examples are not intended to be comprehensive, and there are many other states with similar rules or incentives that are not discussed.

At the time of this writing, twenty-six states have requirements in place for the consideration of externalities in utility decision making. These requirements provide an opportunity for states to include factors that promote the implementation of high efficiency cogeneration, if they so choose.

Thirteen states have quantitative externality requirements, in which five have endorsed specified, monetary values for external environmental costs. These five states are California, Massachusetts, Nevada, New York and Wisconsin. Since these five states have the most specific and quantitative requirements, some discussion of their regulations is important.

## California

In June 1991 the California PUC ordered an investigation into the possibility of including environmental externality costs as a factor in the bidding criteria in Standard Offer No. 4 (CPUC Order 91-06-022). In April 1992 the Commission issued an order (92-04-045) stating that environmental externalities would be used in future decisions to obtain Least-Cost Resource Plans. Costs per ton of emissions were developed, and these costs are to be considered in the cost-effectiveness calculations used in evaluating alternatives in new Resource Plans. The bidding criteria for cogenerators contained in Standard Offer Number 4 have been modified to include these environmental costs relative to alternate utility emission rates in the form of an "adder" or a "subtracter."

If a cogenerator wins a bid, they are then eligible for an added payment if their emissions are below the alternate utility unit—or conversely, a subtracted payment if they have emissions higher than the alternate utility unit.

For an efficient cogenerator there are two ways they can benefit—one is the added payment that they get if they win the bid based on their fixed and variable costs of energy production, and the second is that they could reduce their bid price below their actual fixed plus variable cost of production with the knowledge that they would qualify for the "adder" if they win the bid.

The 1995 FERC order discussed above has a definite effect on this California approach. In that order, FERC specifically stated that "A state, however, may not set avoided cost rates, or otherwise adjust the bids of potential supplies by imposing environmental adders or subtractors that are not based on real costs that would be incurred by utilities. Such practices would result in rates which exceed the incremental costs to the electric utility, and are prohibited by PURPA."

As a result of this ruling the CPUC encouraged San Diego Gas and Electric and Southern California Edison Company and their bidders to have negotiated settlements. Southern California Edison has negotiated at least one settlement based on this CPUC order as of the end of 1995.

However, in general, little is happening in this regard, since the utilities are mostly waiting until new state restructuring plans come out on wholesale production and direct access before making additional commitments. Even though the FERC order struck down the use of the adders and subtractors that were not based on actual costs, there were other alternatives provided by FERC to accomplish the same goal.

**Massachusetts**

In August 1990, Order #89-239 was issued by the Department of Public Utilities (DUP) concerning the integrated resource management process that requires electric companies to use environmental externality values proposed by the Department of Energy Resources in their QF bidding. Order #89-113A and 89-119A (5-31-90) directed Cambridge and Commonwealth Utilities to consult with the Department of Environmental Protection concerning how to incorporate Best Available Control Technology ("BACT") into the Request for Proposal (RFP). BACT assumptions would be summarized for informal purposes for use by RFP respondents in preparing their bids.

Recently, in December 1994 the Massachusetts Supreme Judicial Court struck down the state's environmental externalities rules that would have imposed an environmental adder on the costs of coal-fired power. The court said, "the department does not have responsibility for the protection of the environment..."

As a result of this court order, the DUP issued order# 94-162 requiring utilities to remove externalities as a criteria in the QF bidding process. At the time, this was a set back for encouraging efficient cogeneration. However, in light of the recent FERC ruling, Massachusetts has been provided with a set of future possibilities that they know will meet FERC approval, and may also help satisfy the concerns of the state Supreme Court.

**New York**

In 1989, the New York Public Service Commission ordered (#88-E241, opinion # 89-7) that the weight for environmental considerations be set at 24% of avoided costs in evaluating bid proposals and demand-side management programs. Two-thirds of that amount was attributed to air emissions. This is an incentive for high efficiency cogeneration. New York has not altered their position in light of the recent FERC ruling, and apparently believes that their regulation is in compliance with the FERC rule.

**Nevada**

In January 1991, the Nevada Public Service Commission adopted a rule on resource planning for electric utilities that required an analysis of environmental externalities (Docket No. 89-752). In part, the rule requires a utility's power supply plans to include a statement quantifying the environmental costs and the net economic benefits to the state from each option for future supply.

Section 7 of the final rule requires the utilities to quantify the environmental costs the state would realize from the air emissions associated with implementing and maintaining each plan for supply or demand. If Nevada were to require use of a fuel efficiency factor in the calculation of full avoided costs, this would encourage QF cogeneration facilities to have higher thermal efficiencies and corresponding lower air emissions.

**Wisconsin**

In December 1988, the Wisconsin Public Service Commission decided to conduct a study on whether to continue paying the retail rate to renewable technologies that do not involve combustion or that use renewable resources to generate electricity and may have similar economic and environmental benefits. This decision was part of a docket (Docket

6630-UR102) to consider changes in Wisconsin Electric's net energy billing tariff that would reduce the rate paid to small power producers who were net energy producers.

In 1989, the Commission decided to address the related issue of whether to include a non-combustion credit in the buyback rate for small power producers. As a result of the study, the Commission decided (in Docket 5-EP-6) to exempt renewable resources under 20 kW from having to receive a reduced retail rate. The Wisconsin commission could include efficient cogeneration technology in its study of possible incentives for renewable non-combustion technologies. Although the cogeneration process involves combustion, the dual use of the combustion process to both provide steam for an industrial process and provide electricity should be favored as minimizing combustion.

Eight other states have quantitative requirements in place for the consideration of externalities in utility decision making, but do not assign specific numerical values to these requirements. These states are: Hawaii, Iowa, Montana, New Jersey, Ohio, Oregon, Utah, and Vermont. In addition, Florida is considering adopting a rule that includes an efficiency requirement to qualify for a Standard Offer Contract; and Virginia has also put a cogeneration incentive in place. Five of these state regulations and policies are discussed below:

### Florida
The State of Florida Public Service Commission is setting up a Standard Offer for cogeneration power purchasing that contains an efficiency requirement. The FPSC Proposed Staff Rule went to the full Commission in January, 1996. The Proposed Rule defines a High Efficiency Cogenerator as one where the sum of the electrical output in Btus and the thermal output in Btus divided by the total Btu input to the plant must be greater than 75% of the fossil fuel input.

For a cogeneration facility that is entirely fossil fuel powered, this is equivalent to a minimum 75% UTEP standard. If this Proposed Rule is accepted by the Florida Commission, the result will be a Standard Offer that will encourage the construction of high efficiency cogeneration units.

### New Jersey
The New Jersey Board of Public Utilities, (Docket No. 8010-687B), set up a Procurement Evaluation System from 1988 to 1993 in which they

established a ranking system for bid evaluation by utilities and gave a minimum weight factor of 20% to non-economic factors. This weight factor was intended to promote purchases from cogeneration and small power production facilities.

The Board said that the utilities' evaluation criteria must include both environmental benefits and fuel efficiency. This settlement stipulation was in effect from 1988 to 1993, but was not continued after that date. Although this settlement stipulation did not directly address the issue of cogeneration efficiency, New Jersey could have promoted efficient cogeneration if fuel efficiency were explicitly listed as one of the non-economic factors.

### Oregon

In its 1993 Least Cost Planning Order (93-695), the Oregon Public Utility Commission required that external environmental costs be considered in evaluating the cost effectiveness of resource options. The commission developed a competitive bidding system in which environmental degradation would be a criteria of at least 10%, and thus encourage efficient cogeneration from QFs.

### Vermont

In 1990, the Vermont Public Service Commission required an investigation of conservation and energy efficiency in the LCUP. The Commission stressed a need for cost-effective investments in energy efficiency to improve the environment. The PSC ordered utilities to increase their supply-side costs by 5% in resource planning in order to capture certain costs such as pollution that were not already included in the calculated prices of fossil fuel supply sources. The Vermont PSC could provide additional incentives for investments in fuel efficient cogeneration.

### Virginia

The state of Virginia has also looked at the issue of cogeneration efficiency. In June, 1990, the Virginia State Corporation Commission issued a staff report inviting comments on the issue of promoting highly efficient cogeneration projects as part of the rule development process for competitive bidding programs. The staff proposed that the efficiencies associated with cogeneration should be given some weight as a non-price factor in the evaluation of competitive bids from QFs.

The Commission did not agree because it thought that bids from efficient cogenerators would already reflect their lower fuel costs. The Commission did decide that the lower emissions from efficient cogenerators could be considered as a viable factor in the assessment of costs for environmental externalities, and ordered utilities to increase avoided cost payments to small cogeneration facilities by 15% to account for environmental benefits.

## POLLUTION TAXES

States could set pollution tax rates on a pollutant per kilowatt hour generated basis. Efficient cogenerators would pay less tax per kilowatt-hour and would realize higher profits from the sale of their electricity.

## OTHER RELATED STATE POLICIES
## WHICH COULD ENCOURAGE COGENERATION

A number of options are available to states to circumvent the low FERC standard for cogeneration efficiency. The most direct is to adopt a higher standard. Others include:

- Differentiated rates on standard rate contracts to encourage efficiency in cogeneration, similar to the FERC rule on different standard contract rates on supply technologies.

- Buyback rates tiered to favor the most efficient cogenerators.

- Wheeling and interconnection costs designed to favor the most efficient cogenerators and thus give them a broader market in which they can compete.

- Bidded buyback rates for qualifying facilities (QFs) which include environmental degradation.

- A requirement for a minimum heat rate, lower than a central-station, coal-fired power plant as part of the qualifying criteria of a

QF. This would ensure no QF will pollute more than a central-station power plant.

# CONCLUSION

Achieving the benefits of cogeneration is possible, but the present regulatory system has subverted the industry instead of promoting it. Federal policy-makers should reexamine PURPA; but in absence of Federal action, individual states should take the initiative to encourage the building of efficient cogeneration. This chapter has identified a wide range of options that states can pursue to encourage high efficiency cogeneration if they choose to do so.

*Chapter 23*

# Cogeneration As A Retrofit Strategy

*Chapter 23*

# Cogeneration As A Retrofit Strategy

*Milton Meckler, P.E., AIC*
*President*
*The Meckler Group*

## COST ANALYSIS

A cost analysis will often determine that operating costs for a cogeneration system are lower than those for a system using purchased energy. The first step in a complete analysis is to compile an annual cash flow list for every year of the projected life of a proposed system (Table 23-1 shows a summary of annual costs for a typical application).

The estimated useful life of cogeneration systems using gas engines or turbines ranges from 20 to 25 years. Systems using steam turbines have an estimated useful life to 30 to 35 years. It has been proven that within limits and appropriate conditions, cogeneration will save money and costly fuel. Yet, due to natural restrictions in conversion of fuel for conventional power use, less than 40% of this energy is converted into power. The remaining "waste heat" is left unused during the generation of electricity or operation of mechanical equipment.

The use of waste heat in processes other than power generation increases the amount of fuel energy available. Still, not every power generation facility is able to put this additional energy source to use. However, many industrial, commercial and institutional facilities (in and outside of the U.S.) utilize this valuable energy saving commodity. In-

**Table 23-11. Summary of Annual Costs for High-Pressure Boiler with Steam Turbine**

HIGH PRESSURE BOILER WITH STEAM TURBINE

INITIAL INVESTMENT: $576,000

| Year | Rate $/kWh | ANNUAL ENERGY USAGE | | | Maintenance & Operation $/yr | Cost Sub-total $ | Capital Recovery $ | Total Annual $ |
| | | Purchased Electricity $/Yr | Fuel Cost @ 3.90/MCF $ | Water $/yr | | | | |
|---|---|---|---|---|---|---|---|---|
| 1980 | | 1,302,056 | 692,250 | 41,800 | 187,000 | 2,223,106 | 63,160 | 2,286,266 |
| 1981 | | 1,540,766 | 865,312 | 45,980 | 205,700 | 2,657,758 | 63,160 | 2,720,918 |
| 1982 | | 1,822,878 | 1,081,640 | 50,580 | 226,270 | 3,181,368 | 63,160 | 3,244,528 |
| 1983 | | 2,085,459 | 1,352,050 | 55,640 | 248,897 | 3,742,046 | 63,160 | 3,805,206 |
| 1984 | | 2,397,952 | 1,690,063 | 61,200 | 273,787 | 4,423,002 | 63,160 | 4,486,162 |
| 1985 | | 2,758,188 | 1,842,169 | 67,320 | 301,165 | 4,968,842 | 63,160 | 5,032,002 |
| 1986 | | 3,005,579 | 2,007,964 | 74,050 | 331,282 | 5,418,875 | 63,160 | 5,482,035 |
| 1987 | | 3,276,840 | 2,188,681 | 81,460 | 364,410 | 5,911,391 | 63,160 | 5,974,551 |
| 1988 | | 3,569,803 | 2,385,662 | 89,000 | 400,851 | 6,445,916 | 63,160 | 6,509,076 |
| 1989 | | 3,893,147 | 2,600,372 | 98,560 | 440,936 | 7,033,015 | 63,160 | 7,096,175 |
| 1990 | | 4,242,532 | 2,834,405 | 108,420 | 485,030 | 7,670,387 | 63,160 | 7,733,547 |
| 1991 | | 4,624,468 | 3,089,502 | 119,260 | 533,533 | 8,366,763 | 63,160 | 8,429,923 |
| 1992 | | 5,041,126 | 3,367,557 | 131,190 | 586,886 | 9,126,759 | 63,160 | 9,189,919 |
| 1993 | | 5,494,675 | 3,670,637 | 144,300 | 645,575 | 9,955,187 | 63,160 | 10,018,347 |
| 1994 | | 5,989,457 | 4,000,994 | 158,740 | 710,132 | 10,859,323 | 63,160 | 10,922,483 |
| 1995 | | 6,527,640 | 4,361,084 | 174,600 | 781,145 | 11,844,469 | 63,160 | 11,907,629 |
| 1996 | | 7,115,735 | 4,753,582 | 192,470 | 859,260 | 12,921,047 | 63,160 | 12,984,207 |
| 1997 | | 7,755,913 | 5,181,404 | 211,280 | 945,186 | 14,093,783 | 63,160 | 14,156,943 |
| 1998 | | 8,452,513 | 5,647,730 | 232,400 | 1,039,705 | 15,372,348 | 63,160 | 15,435,508 |
| 1999 | | 9,214,215 | 6,156,026 | 255,640 | 1,143,675 | 16,769,556 | 63,160 | 16,832,716 |
| 2000 | | 10,043,191 | 6,710,068 | 281,210 | 1,258,042 | 18,292,511 | 63,160 | 18,355,671 |
| TOTAL | | 100,154,133 | 66,479,152 | 2,675,700 | 11,968,467 | 181,277,452 | 1,326,360 | 182,603,812 |

NOTE: ELECTRICAL ESCALATION AS FOLLOWS: 18% FOR THE YEARS 1980-81, 15% FOR THE YEARS 1982-83 AND 9% THEREAFTER. NATURAL GAS ESCALATION AS FOLLOWS: 25% FOR THE YEARS 1980-84, 9% THEREAFTER. CAPITAL RECOVERY FACTOR @ 9%/YEAR FOR 20 YEARS. ANNUAL FUEL CONSUMPTION: 177.500 × 10⁶ CU-FT.

stallation of energy consuming equipment that relies on heat-installed power (i.e., absorption cooling rather than mechanical cooling) is common. It represents a dual advantage by simultaneously decreasing overall power requirements and increasing the use of waste heat, while improving control of electrical utility channels. Waste heat recovery can improve cost and energy savings, but these savings are available only under the proper conditions. A cogeneration feasibility analysis is needed to determine these conditions for each facility.

# FEASIBILITY ANALYSIS

The cogeneration feasibility analysis involves three distinct phases: (a) data collection, (b) analysis, and (c) report documentation. Developing a meaningful analysis is a complex process. It requires a working knowledge of sound engineering and economic principles, orderly procedures and calculations. A broad perspective, good judgment and objectivity also are required.

Important factors of the overall feasibility analysis include technical and economic analyses. The technical analysis consists of determining energy characteristics and consumption to select suitable equipment types and sizes to meet system requirements. For example, single waste heat recovery is a primary function of cogeneration prime movers, waste heat boilers, and associated recovery equipment that requires special attention.

Consideration should be given to alternative methods of supplying energy needs (mechanical vs. absorption cooling, or electric motors vs. prime mover direct drives, for instance). This is necessary to optimize the demand balance between prime movers and waste heat recovery. After a technically feasible system is selected, the equipment performance characteristics can be applied to its energy load requirement. This will determine primary energy consumption such as fuel or electricity, usually for a period of one year.

The economic analysis helps to estimate the initial investment and annual operating cost of each alternative system analyzed in the technical analysis. Cost estimates are developed by suitable procedures in order to determine the most economically feasible system.

## Cost Factor Considerations

Assuming that cogeneration packages become more standardized for commercial applications, most of the high costs associated with today's cogeneration market will be alleviated. Still, research and development work is needed to reduce specific costs such as those related to connecting cogenerators with utility grids.

For utilities, a major concern is that a great number of small cogeneration systems connected to the grid may deteriorate the level of safety maintenance on the grid. Their aim is to maintain high power quality, prevent variations in load, current and voltage, and to protect powerline workers from hazards.

Several interconnection guidelines can be used for safe operation. Protective relaying is an important requirement for operating a cogeneration system in parallel with the grid, and it also protects the cogeneration system.

There is, however, a lack of published technical data on interactions between the grid and cogeneration hardware, making the policies regarding interconnect practice inconsistent. These uncertainties are causes of unnecessary costs and delays for cogeneration development.

Another barrier limiting applications for packaged cogeneration is the high cost of installation. In California, for example, installation cost is often 50% of the total commercial cogeneration system costs. An interesting research program now being conducted, aimed at reducing installation costs, involves the direct integration of cogeneration systems with commercial rooftop units. This work is based on valid foresight which predicts that the future of cogeneration in the retrofit market (i.e., for application to restaurants, hotels and other commercial buildings) depends on inexpensive integration with packaged air-conditioning units generally found in commercial applications.

The concept developed by Gas Research Institute (GRI) of Chicago involves a patented heat transport system utilizing a refrigerant to couple the direct expansion (DX) coil of existing rooftop HVAC equipment and the absorption chiller of a cogeneration package. A direct refrigerant interface takes away the need for water coils, pumps, and other expensive modifications. This concept has been tested successfully with common rooftop units and has positive results for the technical feasibility of retrofitting commercial HVAC systems with cogenerated space cooling and heating.

# PRIME MOVERS

Because of the difference between electrical and steam loads, two methods are used to size equipment for a cogeneration plant. These are: (a) selecting equipment best suited to supply electrical demand and supplementary boiler firing when steam demand is higher than generated steam, and (b) choosing equipment to meet steam demand without supplementary boiler firing.

Reciprocating engines are probably the most widely used prime movers due to their comparable low-cost and various ranges of size and speed. They can be fueled by oil, natural gas, liquid propane gas, digester gas, or a combination of these fuels. Dual-fuel engines which can switch fuels while running offer a higher degree of reliability.

Complete "total energy" modules consist of an engine generator unit, heat recovery unit, switchgear and controls. These units lower equipment and installation costs, decrease construction time, and place the burden of responsibility on one supplier. Generator ratings normally are based on continuous duty operations at a power factor of 0.8. The most common size and speed engine generators used in cogeneration systems, are shown in Table 23-2.

**Table 23-2. Common Generator Sizes and Speeds.**

| kW range | Rpm | Cycle | Fuel |
|----------|-----|-------|------|
| 150-800 | 1,200 | 4 | Oil, gas, LPG |
| 450-3,000 | 720 or 900 | 2 and 4 | Oil, gas or dual fuel |
| 1,650-5,000 | 514 | 2 | Dual fuel |

Factors favoring gas turbines as prime movers include lightweight, minimum foundation requirements, rapid start-up capability, and automatic operation with minimum supervision. Unfavorable factors include a thermal efficiency generally less than reciprocating engines, rapidly falling efficiency at part load for the single-shaft gas turbines, and substantial loss of power output with higher ambient temperatures and high initial power cost.

## Gas Turbine Categories

Based on shaft arrangement, gas turbines fall into two categories: single-shaft and two-shaft. The single-shaft units are more suitable for constant speed applications such as generators. Their design is mechanically simpler for reducing cost. Also, their ability to handle speed variations is excellent with simple fuel control. Two-shaft gas turbines permit a wide variation of full power speeds.

There also is an inherent lag in response between the gas producers and power turbines, though each can be equipped with a speed governor. Only the gas producer turbine requires a starter to accelerate and pressurize the system. Gases leaving the gas turbine flow through the free power turbine forming a fluid coupling. The remaining energy is absorbed by the power turbine and transferred to the output shaft which drives the load of the electric generator.

Turbine speeds vary from 40,000 rpm to 60,000 rpm for 50 hp to 2,000 hp (37 kW to 1490 kW) to as low as 3,600 rpm for 20,000 hp to 30,000 hp (14,920 kW to 22,380 kW). Starter systems include compressed air, gas starter (popular in the U.S.), ac/dc motors, batteries, chargers, motor-generator sets, reciprocating engines, high-pressure impingement, and hydraulic and hand cranks. Some turbines burn two types of fuels, usually gas and oil, incorporating two complete fuel systems that could be switched while in operation.

Gas turbines meet required steam and electrical demands ranging from 21,160 pounds per hour (lb/h) at 4 MW to 25,000 lb/h at 4.8 MW. They can be designed to meet demands with or without supplementary boiler firing, depending on the capacity range of operation and the number of individual turbines used. Steam turbines, either condensing or non-condensing, are used for many industrial applications. Mechanical drive steam turbines are ideally suited to drive mechanical equipment such as refrigeration compressors, pumps, fans, and other plant auxiliaries. They are efficient and economical. Monetary considerations tend to group turbines in terms of pressure, temperature, and steam flow, as indicated in the pressure and temperature classes shown in Table 23-3.

### Back-pressure Turbines

Most mechanical drive turbines in process industries are either back-pressure or reducing-valve types. An assured demand for low-pressure steam in a central plant can make back-pressure turbines very attractive to use.

**Table 23-3. Turbine Types (Pressure and Temperature Classes).**

| Unit size | Pressure (psig) | Temperature °F |
|-----------|-----------------|----------------|
| Small     | 150-400         | 500-750        |
| Medium    | 400-600         | 750-825        |
| Large     | 600-900         | 750-950        |
| Large     | 900-2,000       | 825-1,050      |

In self-generation, the cost of high pressure boilers often can be justified. This entails using a high-pressure, multistage turbine exhausting into a medium-pressure turbine drive for a centrifugal refrigeration machine and exhausting this turbine into an absorption machine at low pressure. Due to high steam rates, back-pressure turbines must be highly efficient to fit into the heat balance of an existing plant. Therefore, multistage turbines are required. A back-pressure turbine is installed with an exhaust pressure regulator to provide constant exhaust pressure.

**Condensing Turbines**

In a condensing turbine, steam expands from throttle conditions to a condensing pressure. Its main advantage is a low steam rate. The higher the pressure at inlet conditions, the lower the steam rate for the same condensing pressure. Throttle conditions are constant and determined by economics, except in some process industries where there is an abundance of waste steam.

A single-stage turbine is economically attractive in small sizes. However, additional equipment is necessary to implement the condensing cycle. This equipment may include a surface condenser, hot well pumps, steam ejectors or vacuum pumps to maintain condenser vacuum, a cooling water system involving cooling towers, and other accessories which increase initial costs.

Condensing turbines are best suited when the demand for process steam is not simultaneously required with power generation or shaft output, since steam would then become a waste product. When available, waste steam cannot be used by the condensing turbine. A bypass, however, can be used to conduct it directly to the surface condenser.

## Extraction Turbines

Extraction turbines consist of two turbines in series enclosed within the same casing. Extraction steam serves process needs, or feeds to auxiliary equipment at one pressure. The second turbine, if non-condensing, may serve other process needs at another pressure. A second condensing turbine drives an electric generator.

An extraction turbine may be either the condensing or non-condensing type. In an extraction turbine, all of the steam works up to the extraction point. This portion of the work is done at a higher water rate. The amount that proceeds to the condenser is the only part of the steam that has done work at a low steam rate.

In many central plants, it is possible to use non-condensing extraction turbines for power generation with an exhaust steam extraction point for hot water heating, absorption refrigeration, or other needs. With condensing turbines, an extraction point can feed refrigeration machines. When it is not needed, the steam can proceed to the condensing section to generate power. An extraction turbine is not built for less than 500 hp.

## Single-Stage Turbines

A single-stage turbine is one of the most basic prime movers in industrial plant operations. It is generally considered the "workhorse" of the industry due to its inexpensive drives and design flexibility (depending on first cost for intermediate or low efficiencies).

The relatively high steam rate of a single-stage turbine makes it very suitable for non-condensing service, and its exhaust steam has various uses in process work. Because of its high range (15 hp to 3,500 hp or 11 kW to 2610 kW), it can be used successfully in central plant work for auxiliary pumps, fans, etc., in a centrifugal absorption arrangement. As a condensing turbine, the single-stage turbine is at more of a disadvantage than the multistage turbine due to its higher efficiencies, which can keep steam rates low and steam plant size within reasonable limits.

At certain times, there may be more energy available than needed. The logical solution is to consider selling it back to the utility company. This is a complex activity that must be considered before building or modifying the cogeneration system. Two basic approaches can be used: (a) a cogeneration system running parallel with utility systems, or (b) separate cogeneration systems capable of transferring loads back and forth (i.e., import and export).

If the cogeneration system is separate (that is, it cannot connect in parallel to the utility system), a switching scheme will be necessary (refer to Figure 23-1). A parallel scheme will be easier to operate because the connection between the grid and the cogeneration system is direct (refer to Figure 23-2). With this system, however, the cogeneration scheme must meet utility and government standards.

If cogeneration capability is less than 20 kW, this involves installing controls and devices for suitable connection to the utility. In this case, the devices are required only if the cogeneration system maintains its output when disconnected from the grid.

With this scheme, voltage regulation equipment also will be required. Designs for all of these items must be approved by the utility. Low- or high-voltage surges are not allowed, and the cogenerator is responsible for any damage to the utility or other user equipment. Co-

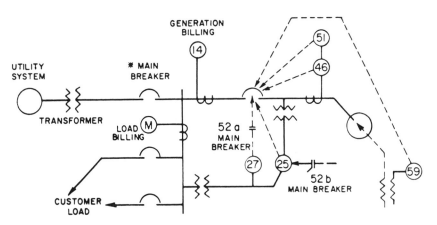

*Synchronizing equipment required if customer desires to close breaker with both sides energized.

**Relay Identification**
25- Synchronizing
27- Undervoltage
46- Phase-Balance
51- Phase-Overcurrent
52- Main Breaker Auxiliary Switch
59- Neutral Overvoltage
M- Meter

**Figure 23-1. Cogeneration Switching Scheme**

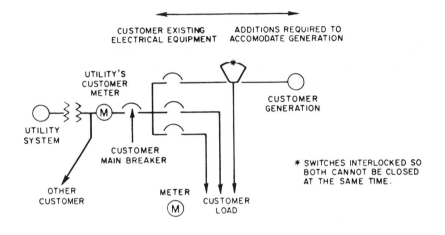

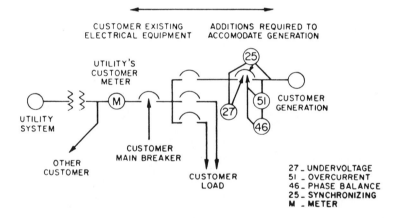

Figure 23-2. Parallel Switching Scheme

generation systems with a capacity of more than 20 kW will require various trip devices, detectors, special controls and assorted interlocks. In addition, cogeneration systems with outputs high enough to affect utility generation or voltage regulation must have monitoring and control equipment that can be remotely operated by the utility. With systems of this magnitude, the utility will have priority over cogeneration systems including start-up, shutdown and output operations.

The cogeneration system operators also must be aware of operating requirements for these larger systems and must notify the utility before any operation. And, the cogeneration system must comply with utility requests for maintenance work shutdown and/or emergency situations. Weekly, bi-weekly and monthly records should be kept for maintenance reporting procedures.

For example, a one-week record on engines will indicate the need for cleaning filters and other engine maintenance. These records also should indicate any overlapping of operating schedules.

Ultimately, all complete data (i.e., log sheets) should be systematically filed to maintain continuous records. These records will help in preparing an effective preventive maintenance program which is essential for dynamic equipment use in the overall system.

# INDUCTION GENERATION

When operated parallel with a utility, induction generators must have a frequency greater than electrical bus frequency (that point in a system where we have nominal voltage and frequency). This will avoid a slip between the generator and bus.

Important differences between a synchronous generator and induction generator are:

(a) A synchronous generator can operate separately from the utility. The induction generator draws its excitation from the bus. When other generators in the utility system fail, the induction generator cannot operate.

(b) A synchronous generator can improve (raise) plant power factor by carrying a reactive load. Since the induction generator takes its excitation from the power line in the form of reactive power, it has a tendency to lower the overall plant power factor, unless corrected. Lower power factor will increase the power bill.

(c) When going on-line, the induction generator and the bus do not have to be as close in frequency and phase as in the case of a synchronous generator. Therefore, paralleling the induction generator with the utility is easier. This is accomplished simply by closing the circuit breaker.

(d)  Induction generators require fewer controls—often just a circuit breaker and simplified relay protection. Controls for a synchronous generator include a regulator, reverse-power relays, voltage control, frequency control, differential and loss of excitation protection.

The generator enclosure may be of the open or totally enclosed water/air-cooled (TEWAC) type. For the TEWAC type, air inside the generator is isolated from the ambient air so that the dust and moisture cannot enter. A fan moves the air inside the generator through windings. Heat is removed from the air by a water-cooled heat exchanger. TEWAC generators are used for outdoor installations, dusty locations and applications requiring minimum noise and attractive equipment. TEWAC generators also provide better heat removal than open generators.

Generators can be built with a variety of internal and terminal connections. Some available electrical connections include delta connection, star (or wye) connection, line-to-line, line-to-ground, 3-wire, 4-wire, 6-wire and 7-wire. The type of connection used will not affect the requirements of the generator driver and will have minimum effect on its controls.

Stator temperature detectors (STDs) are devices embedded in the stator windings to detect the temperature windings. Two per phase can be supplied if stator resistor temperature detectors (RTDs) are specified. If one RTD in a phase fails, other RTDs can be used since replacing a stator RTD is a major repair job.

Generator controls are needed to connect the generator to the customer electric power system. Generally, the following items are included in the generator control unit:

(a)  **Circuit breaker:** This switch is used to manually connect the generator to the line or take the generator off the line (either manually or automatically). The circuit breakers are sized to trip at circuit demand. They are calibrated 15% above the required current to produce the rated kVa at a rated voltage.

(b)  **Shunt trip device:** This causes the circuit breaker to trip in response to an electrical signal. The turbine shuts down in response to a safety device. A switch on the turbine will send an electrical signal to the shunt trip, which causes the generator to go off-line. Some devices that will trip the unit are:

- overcurrent relay,
- generator differential relay,
- thermal overload relay,
- low lube oil pressure,
- transformer sudden pressure relay,
- undervoltage,
- main and bust tie circuit breakers open, and
- control switch.

(c) **Ac ammeter:** This device indicates the output current and the generator input current to the motor.

(d) **Ac voltmeter:** Indicates the bus voltage.

(e) **Wattmeter:** Indicates the real component of the generator's electric power output or input to the motor. This meter is zero centered.

(f) **Recording wattmeter:** Same as wattmeter.

(g) **Current transformers and potential transformers:** Reduce current/voltage from the generator output terminals to proportionally lower values suitable for instrumentation. These are furnished as needed for the 3-phase instruments in the generator control unit.

(h) **Bs voltmeter:** Indicates bus voltage.

(i) **Governor control switch:** Controls the motor-operated speed changer. It also controls the speed setting of the governor, thus controlling the generator load.

(j) **Reverse power relay:** Indicates electric power flowing from (or to) the bus into the generator. It is needed to show whether the generator is in the generating or motoring mode.

(k) **Watt-hour meter:** Summarizes the total amount of energy delivered to the system.

To illustrate the feasibility, analysis and benefits of a cogeneration system, the following case history of an institutional-industrial application is summarized. High temperature water (HTW) production, quality and reliability were the facility's most important requirements. Because of the critical nature of production losses, continuous electrical and gas services were essential. Cogeneration, fired by natural gas with back-up systems, offered the facility assurance of a reliable source of electrical power and thermal energy.

For each alternative, this analysis required evaluation of the estimated electrical power production, gas and energy consumptions (compared to the existing system). An economic analysis was needed to indicate the most cost-effective scenario and option. The options analyzed were all gas turbine cogeneration scenarios with different applications and modes of operation. A quick review of these options is as follows:

1.  **Existing facility:** Three HTW generators, conventionally fired to produce hot water at 400°F. Hot water would be used as the heat transport medium to provide heating for all facilities where required.

2.  **Scenario I, Option 1:** Three gas turbines operating at constant full load. The existing three HTW generators would be used as heat-recovery heaters. Excess hot water would produce steam to operate a small steam turbine to generate additional electricity.

3.  **Scenario I, Option 2:** One large turbine operating at full load with the turbine exhaust equally ducted to each of the three existing HTW generators. Excess heat would be passed through a heat exchanger to produce steam to operate a small steam turbine to generate additional electricity.

4.  **Scenario I, Option 3:** One large gas turbine operating at full load with turbine exhaust passing through a new waste heat boiler sized to match the turbine. The other existing boilers would be used as back-up units. The steam turbine cycle was included, and would generate additional electricity.

5.  **Scenario II:** Three gas turbines, each coupled to one of the three existing HTW generators. Each turbine would vary its load and alternately start and stop in accordance with the thermal demand. Therefore, no excess heat would be generated and a steam turbine cycle would not be needed.

6. **Scenario III, Option 1:** One gas turbine sized to meet minimum thermal demand, operating continuously. One of the three existing boilers would be used as a waste heat boiler. Any thermal load in excess of what is provided by the turbine exhaust would be supplied by conventionally firing the other boiler(s).

7. **Scenario III, Option 2:** The basic operation is the same as Scenario III, Option 1. Instead of firing another boiler, however, supplemental firing would be done by directly firing the turbine gases exhausted to the hot water generators.

All scenarios were analyzed to determine gas consumption, generated electrical power and the amount of heat recovery expected from each option (refer to Figures 23-3 through 23-5). New domestic hot water (DHW) systems were calculated for three major areas: the cafeteria, maintenance and the calibration building.

The cafeteria required about 116 gallons/hour maximum demand capability, which required a maximum heating capacity of 96,300 Btuh. Normal demand would probably be reached very sporadically. The maintenance building's maximum demand was calculated to be 770.4 gallons/hour. This also was the storage tank size. To meet demand, a heating capacity of 624,000 Btuh was needed.

This system was more expensive and required larger runs of pipe and facility area expansions in addition to the cost and installation of the heaters and concomitant equipment. Maximum DHW demand for the calibration building was calculated as 30 gallons.

A 40-gallon storage tank, however, was recommended because of its standard size and only slightly higher cost. The heating demand was calculated to be 4,265 Btuh. Existing HTW boilers operated with a maximum efficiency of 78.3% at a full-load of $25 \times 10^6$ Btuh each. Operating at a minimum pressure of 250 psig, the generators were capable of producing hot water at 400°F.

The boilers had dual fuel capability with natural gas as the primary fuel, and #2 fuel oil as the standby fuel source. During boiler plant operation from April to November, only one boiler was operated. Two boilers were operated from November to April. The boilers were instrumented so that the total load was distributed equally between the two boilers. A boiler plant operator was present on a 24-hour basis throughout the year.

## 3 - 800 KW GAS TURBINE GENERATORS EXHAUSTING TO 3 HEAT RECOVERY HOT WATER GENERATOR UTILIZING STEAM GENERATOR AND SUPPLYING DOMESTIC HOT WATER SCHEMATIC. (SCENARIO I, OPTION I)

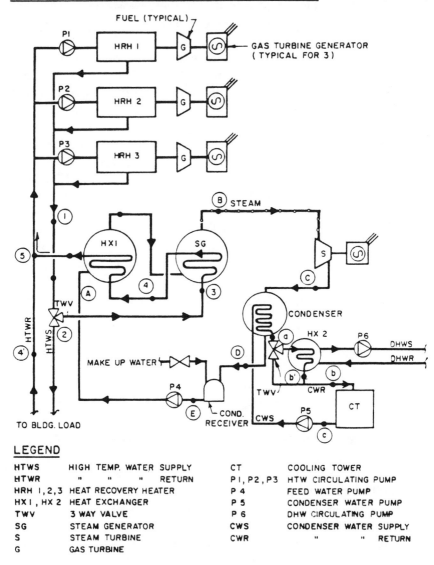

### LEGEND

| | | | |
|---|---|---|---|
| HTWS | HIGH TEMP. WATER SUPPLY | CT | COOLING TOWER |
| HTWR | "     "     " RETURN | P1, P2, P3 | HTW CIRCULATING PUMP |
| HRH 1,2,3 | HEAT RECOVERY HEATER | P 4 | FEED WATER PUMP |
| HX1, HX 2 | HEAT EXCHANGER | P 5 | CONDENSER WATER PUMP |
| TWV | 3 WAY VALVE | P 6 | DHW CIRCULATING PUMP |
| SG | STEAM GENERATOR | CWS | CONDENSER WATER SUPPLY |
| S | STEAM TURBINE | CWR | "     " RETURN |
| G | GAS TURBINE | | |

Figure 23-3. Scenario I, Option 1.

## - 3200 KW GAS TURBINE GENERATOR EXHAUSTING TO 3 HEAT RECOVERY GENERATORS UTILIZING A STEAM TURBINE GENERATOR AND DOMESTIC HOT WATER HEAT EXCHANGER SCHEMATIC.

(SCENARIO I, OPTION 2)

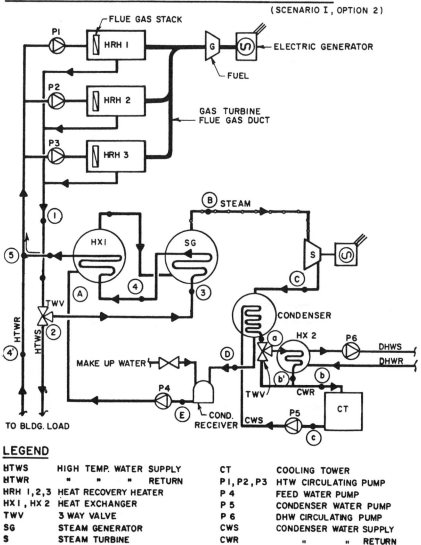

Figure 23-4. Scenario I, Option 2.

### LEGEND

| | | | | |
|---|---|---|---|---|
| HTWS | HIGH TEMP. WATER SUPPLY | | CT | COOLING TOWER |
| HTWR | " " " RETURN | | P1, P2, P3 | HTW CIRCULATING PUMP |
| HRH 1,2,3 | HEAT RECOVERY HEATER | | P 4 | FEED WATER PUMP |
| HX 1, HX 2 | HEAT EXCHANGER | | P 5 | CONDENSER WATER PUMP |
| TWV | 3 WAY VALVE | | P 6 | DHW CIRCULATING PUMP |
| SG | STEAM GENERATOR | | CWS | CONDENSER WATER SUPPLY |
| S | STEAM TURBINE | | CWR | " " RETURN |
| G | GAS TURBINE | | | |

## I-3200 KW GAS TURBINE GENERATOR EXHAUSTING TO A HEAT RECOVERY HOT WATER GENERATOR UTILIZING A STEAM TURBINE GENERATOR AND DOMESTIC HOT WATER HEAT EXCHANGER SCHEMATIC.

( SCENARIO I , OPTION 3 )

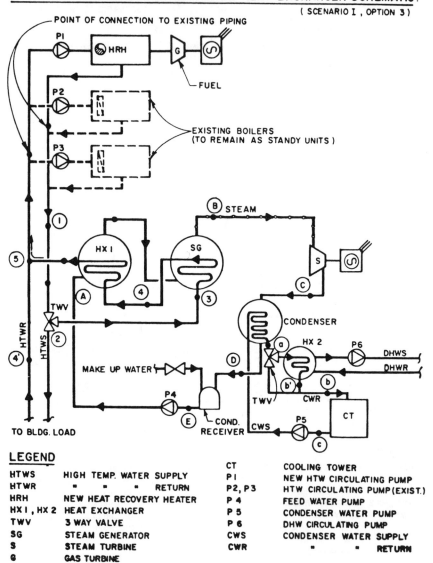

**LEGEND**

| | | | |
|---|---|---|---|
| HTWS | HIGH TEMP. WATER SUPPLY | CT | COOLING TOWER |
| HTWR | "    "    " RETURN | P I | NEW HTW CIRCULATING PUMP |
| HRH | NEW HEAT RECOVERY HEATER | P2, P3 | HTW CIRCULATING PUMP (EXIST.) |
| HX I , HX 2 | HEAT EXCHANGER | P 4 | FEED WATER PUMP |
| TWV | 3 WAY VALVE | P 5 | CONDENSER WATER PUMP |
| SG | STEAM GENERATOR | P 6 | DHW CIRCULATING PUMP |
| S | STEAM TURBINE | CWS | CONDENSER WATER SUPPLY |
| G | GAS TURBINE | CWR | "    " RETURN |

Figure 23-5. Scenario I, Option 3.

HTW was distributed to the facility by a high-pressure piping system. It provided for all domestic hot water and space heating requirements as well as all heating processes required by the building's maintenance facilities. HTW demand was based on actual boiler logs from 1979. Boiler load data were averaged for each month, and the day of the month closest to the average became a typical day. This was done separately for weekdays and weekends. The typical day was based on thermal loading rather than electrical demand.

The electrical system at the facility was serviced by 34.5-kV loop feeders. The 12-kV feeder system was a combined overhead and underground system. Throughout the base, the 12-kV is transformed to use voltages at various substations. The intent was to interface with the existing 12-kV system at the cogeneration plant site. This would allow base flexibility in operation of the cogeneration plant.

# DEVELOPING LOAD PROFILES

Various electrical load profiles were developed from computer print-outs for the previously chosen typical weekday and weekend conditions for each month. Figure 23-6 shows this monthly electrical consumption of the facility for a typical year. Thermal and electrical profiles were determined for each month. The peak thermal demand indicated in December was significant; the cogeneration system must have the capability of meeting the load. Therefore, the thermal load dictated the electrical power generation.

Since the primary concern of the facility was to provide the necessary thermal load, the electrical generator had to be sized coincident with the turbine exhaust system that could best meet the load. The monthly

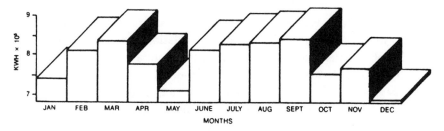

**Figure 23-6. Monthly Electrical Consumption**

electrical energy use varied from 693,000 kWh/month to 845,100 kWh/ month. The peak demand was 4,234 kW.

An analysis of the thermal demand showed wide seasonal variations, unlike the average monthly electrical demand profiles which did not vary significantly. The HTW load showed seasonal fluctuations. Demand profiles for weekday operations were similar for every month. Electrical demand profiles had a low of 400 kW to 600 kW and a high of about 2,400 kW. These occurred at about the same time of the day.

Thermal load analysis was performed using the previous operating year data contained in log books obtained from the facility boiler plant. The electrical load analysis was based on data obtained from the serving facility and analyzed for demand and consumption. Economic analyses were performed on a 25-year life-cycle basis beginning in 1985. Different scenarios of the total life-cycle costs were compared to the existing facility's total cost, then compared to each other on the basis of savings-to-investment.

Cogeneration provided an effective energy conservation and cost-saving technology for the facility. Furthermore, a cogeneration system, fired by natural gas with the existing system, was used as back-up. It offered additional protection against production losses if utility electrical service is interrupted. Accordingly, Scenario I, Option 3 was determined to be the most cost-effective scenario and chosen for implementation (refer to Figure 23-5).

## OPTIONS AND RESEARCH

Recent widespread interest in cogeneration, along with numerous system applications available, has raised the question of which cogeneration set-up to use. The most common answer is to use a system of turbines fired by natural gas. Deregulation has affected the power generation industry as well as the nation's gas companies. More and more, electric utilities facing new rules and regulations are considering the benefits of cogeneration in serving both the company and its clients.

In an effort to manage electricity demand in areas where generation capacity is strained, some utilities are considering gas-fueled cogeneration as a reliable source for the future. The potential for expansion of cogeneration for commercial application (less than 1,000 kW) is being actively developed since large-scale systems have proven reliable. More and more, big-name engine manufacturers (as well as smaller companies) are marketing new cogeneration packages. Some of these companies are now geared toward the down size or "microsystem" trend (away from the multi-megawatt scale) to only 30 kW or less.

Packaged systems respond well to needs of the commercial market, yet are still expensive on a dollars-per-kilowatt basis. Installation costs of commercial systems are disproportionately high and the market is still not mature. Profitable small cogeneration projects are becoming harder to find. The first commercial ventures were established by third-party developers who would contract the work, provide financing for the user, and in return, get a percentage of the savings in energy costs. This practice has been limited due to a saturated market and smaller energy savings. Thus, to succeed, organizations will need financial strength and staying power.

Currently, the cogeneration market is plagued by unexpectedly high installation costs. Primarily, buildings are not designed the way cogeneration systems would want them to be. Yet, in spite of the temporary handicaps, cogeneration systems are gaining a steadfast place in the market. Many chain organizations, hospitals, hotels, motels and restaurants are using cogeneration systems on a trial basis. Should these systems prove successful, cogeneration could become a standard design option for new and remodeled buildings. Surveys have shown that commercial, industrial and institutional building industries using HVAC equipment have the greatest cogeneration potential for the next decade. Future predictions point to an increase in the use of on-site cogeneration systems during the next five years.

Since 1980, natural gas has been the choice of fuel for more than 65% of all U.S. cogeneration projects—usually large-scale system generators. For smaller systems, the trend is greater, showing a gas usage of almost 90%. In the Gulf Coast region of the U.S., natural gas is in plentiful supply. Cogeneration experts agree that gas turbines offer low equipment cost, high electrical generating capacity and a higher exhaust gas temperature. Accordingly, 60% of cogeneration plants built in the U.S. in the past two years are gas-fired. Yet, some of the earliest and most successful cogeneration plants were those built to use waste fuels such as process gas, petroleum, coke or wood.

A number of other fuels can be used in cogeneration—for example, hydrogen-rich off-gases from the chlor-alkali production. The biggest alternative to natural gas is coal. What makes coal so attractive for cogeneration is the rising popularity of fluidized-bed combustors (FBCs). Sulfur emissions from FBCs can be closely controlled through the injection of lime or limestone into the combustion bed. This may eliminate the need for stack-gas scrubbers. It is an especially important factor in cogeneration planning because air emissions are being monitored.

New cogeneration technology involves the construction of modular cogeneration systems. These systems supply steam or heating water as well as electricity to office buildings, laboratories and other commercial buildings. Cogeneration systems benefit almost all facilities. The strongest development of cogeneration in the U.S. has taken place in light industrial areas such as California and Texas. It also has grown in the heavy industrial mid-Atlantic states and in most of the upper Midwest and New England.

Cogeneration installation capacity in the U.S. is estimated to be 14,000 MW to 16,000 MW at this time. Within the next few years, annual additions should approximate 2,000/3,000 MW (based on current data). Crediting increases in electrical energy costs, the demand for cogeneration in this market is likely to continue. The future of cogeneration is difficult to predict; yet, it is a common belief that the economics of cogeneration can be important to the continued manufacture of products in the U.S.

*Chapter 24*

# How a Utility Uses Cogeneration As a "Value Added" Service

*Chapter 24*

# How a Utility Uses Cogeneration As A "Value Added" Service

*Terrence Kurtz*
*Vice President\**
*Central Illinois Light Company*

C ILCO is headquartered in Peoria, Illinois. It serves 190,000 customers in an area of approximately 4,000 square miles. CILCO is a combination utility; its $454 million of 1993 revenue was comprised of $303 million electric and $151 million gas.

Within the state of Illinois, Central Illinois Light Company has the lowest electric rates. This is an absolutely crucial position to be in, that is, the low cost provider. Under any scenario that may unfold in the future, it is strongly felt that it will take more than competitive rates to maintain our customer's base, but low rates are a key component. Increasingly, our customers are demanding additional "value-added" services that in many cases we are best able to provide.

We recognize, in our corporate business plan, that as a value added service, cogeneration can be an important tool to give our company the flexibility we need to supply customers who have particular energy needs with economical energy. Again, more than just competitive rates will be needed to maintain our customer base, which is becoming increasingly aware of competitive alternatives. Industrial customers everywhere are pushing hard for alternatives in their purchase power decisions.

---

\*Mr. Kurtz is now with PSC Energy Services, California.

# CASE STUDY—MIDWEST GRAIN

In early 1993, one of our major industrial customers announced a significant expansion. Midwest Grain, headquartered in Atchison, Kansas, has a major facility in our service territory. They are a fully integrated producer of vital wheat gluten, and beverage and industrial alcohol. Vital wheat gluten is a food additive used to enhance the nutritional value, appearance, texture, taste and other characteristics of baked and processed foods. Beverage and industrial alcohol are produced as part of the gluten processing operations. The beverage alcohol consists primarily of vodka, gin and whiskey. The industrial alcohol is sold for a variety of uses, one of which is a gasoline additive (ethanol). Over 50 percent of our nation's ethanol is produced in central Illinois.

Midwest Grain's process is energy intensive, using large amounts of steam, electricity, and natural gas. Due to its large energy use, Midwest Grain is located in our service territory next to several other major customers that also are very steam, gas and electricity intensive. Because the plants are contiguous, a single plant ultimately supplying all customers is possible. Steam purchases by these industrial customers are equal in importance to electricity purchases. The economics are such that a large-scale project serving several of these sites could be possible.

Between these contiguous sites, a significant portion of our total system's energy is utilized. The loss of not only the existing load but several of the announced expansions would represent a large reduction in revenue and obviously impact our future. Cogeneration can give us and our customers yet one more option to help compete in an increasingly challenging environment.

As mentioned, Midwest Grain is in the process of expanding their processing facilities. This expansion will result in additional energy needs. In fact, between now and 1998, their expansion will most likely double the amount of their electricity, double the amount of their steam use, and triple the amount of their gas consumption. In that CILCO can provide all three of these energy forms, we certainly have a keen interest in retaining them as a major customer.

There exist several gas and coal steam generators on the site that are currently operated by Midwest Grain, as well as a turbine generator operated by them to fulfill some of their electrical energy needs. Midwest Grain has experience and expertise in being a self-generator. What we at CILCO have discovered in working closer with many of our larger self-generating industrial customers is that they are faced with aging equipment—and it is becoming increasingly difficult to meet the more stringent environmental permitting and operating requirements. In many cases, this is beginning to detract resources from their main product line. In short, it isn't as easy to be a self-generator as it used to be.

## THE FIRST STEPS

Again, as part of our business plan to expand our value-added services, we first approached the customer in early 1993. Immediately, we were surprised to learn that they had already budgeted several million dollars of their $75 million plant expansion to install package steam boilers. In fact, they had been out for bids and were within months of awarding the contract.

In addition, they had been approached by an independent power producer who was interested in installing a "PURPA" machine for the industrial park. This factor—the threat of competition—was a major force in moving CILCO's business plan from words into action. We have an excellent working relationship with Midwest Grain, and we asked them if they'd be receptive to our exploring the possibility of an energy partnership whereby we could meet their total energy needs.

In return, they would be freeing up their present personnel, as well as the capital required for the boiler expansion. We would mutually agree to a 15-year steam contract and an 8-year electric contract whereby we would freeze their electric base rate at current levels and agree not to raise them for the next 8 years.

I might add that our company does have a definite retention strategy for our key customers. This strategy involves our corporate sales team signing up specific kilowatt-hour sale targets to long-term contracts. The length of these contracts has varied from 5 years, all the way up to 8 years.

Obviously, for now, we are focussing on our largest retail accounts.

Several objectives were associated with our initial meetings with Midwest Grain. The first and foremost certainly was to identify the needs of the customer, that is, items such as flow rates, pressures, temperatures, timetable, site locations, as well as tolerance for risk on fuel supply and technical specifications of their needs. Certainly another major purpose was to obtain information or intelligence on our competitor, that is, the independent power producer. Finally, and certainly not least, was to begin building the case for CILCO energy in order to distinguish our product from an IPP's product.

As we proceeded with our discussions, we found that we were able to bring several strengths to the negotiating table that we sometimes tend to overlook. Very early in the discussion, it was apparent to us that the customer appreciated our approaching this project specifically from *their needs perspective*, that is, build a plant to meet their needs versus building a wholesale plant whose by-product may coincidentally meet their needs.

The fact that we are a local company with whom the customer was familiar turned out to be a larger strategic advantage than we had originally thought. By identifying the specific needs of our steam host, it became apparent very early in the discussions that a traditional PURPA machine would not be in order. The need for large quantities of reliable steam was the customer's primary concern.

In fact, they would require a stand-by boiler even if there were no power plant. In other words, this customer has some unique project needs to which we had to respond. Consequently, early in the discussions, we trended away from a combustion turbine/heat recovery steam generator technology and began looking at the more traditional cogeneration technology.

## TIGHT TIME FRAME

In discussions with our customer, it was apparent that they had committed to their plant expansion and, therefore, needed process steam to support their anticipated December 1944 start-up. This timetable certainly played to our advantage as compared to the IPP's in that the IPP could conceivable spend large quantities of time and effort trying to find suitable power sales arrangements which would impact the plant's output, as well as the ability for financing.

Our avoided cost at CILCO is less than 2 cents per kilowatt-hour which certainly deters power sales arrangements. All of these were real risks to the project timetable which we pointed out to the customer for strategic purposes. In addition, the legal and regulatory risks of permitting and retail sales and affiliation concerns were also discussed.

CILCO brought several distinct advantages to the negotiating process, particularly from a customer's perspective.

- CILCO's good reputation with our customers. Midwest Grain knows CILCO and is used to working with us on a daily basis in meeting their needs, whereas an IPP has limited presence in the region.

- CILCO has an additional advantage: we can back up our customer's needs with system power. That is, we have many plants, not just one, and the electrical needs would be met from the grid, thereby eliminating the need for separate stand-by power that an outside developer would have to have.

- CILCO can offer stable prices. We would in essence be using a mix of fuels rather than just gas for our electricity, whereas an independent power producer whose electrical production costs are a direct function of natural gas could be at a disadvantage in the event of gas run-ups.

- A CILCO cogeneration project would offer greater certainty in that it was not just a concept on a drawing board that was in search of wholesale or retail customers for its output.

- We are an investor-owned utility that is heavily regulated within the state of Illinois, which offers easier enforceability. Probably a major dispute with an IPP would involve going to court, thereby not being able to offer a regulatory solution.

- CILCO can offer total service, that is, power, steam, gas, DSM programs, financing alternatives, etc.

The bottom line is that Midwest Grain has had a long-term relationship with CILCO and given the right circumstances, we should be in an excellent position for the relationship to continue.

## PLANNING STARTS

After determining the needs of the customer, we concluded there would be no value added for either CILCO or CILCORP to simply put in a boiler on-site. However, there would be real benefits for both Midwest Grain and CILCO if we could install a higher pressure boiler to be used as an on-site steam-electric plant to meet the steam needs of Midwest Grain. We began to explore the possibilities of financing, building, owning and operating this steam plant which would not only meet the current steam needs, but would provide a reserve steam capacity so as to ensure steam reliability which is of critical importance to the customer's process.

We also determined that we would be able to give a share of the savings from the electricity produced back to the customer from this project. In addition, we would be able to provide a long-term commitment on the price of electricity which allows the customer to benefit from low gas prices while ensuring that if prices would spike, they would not pay more than they would if they were buying electricity from us under current rates.

Because our company has not had a rate increase since 1982, freezing base prices for an additional 8 years was very appealing to the customer. These various developments in the first few months made us re-think our whole approach to customers, such as Midwest Grain. We recognize that a traditional utility approach won't cut it in the competitive marketplace which we both face. Merely doing the same old things better would not guarantee long-range success. We needed to look at doing the same old things *totally differently*.

Recognizing that CILCO was drifting into territories we were not completely familiar with, we hired a consultant to assist us in analyzing the competitive marketplace and assess the issues we faced in competing for this account. Their role was to review our competitive advantages and disadvantages and guide us in preparing our proposals. In addition, they assisted us in screening the various technologies available in the marketplace. Purchasing this expertise was critical to our ultimate success.

# THE ENERGY PARTNERSHIP PROPOSAL

In mid-1993, CILCO submitted an Energy Partnership Proposal for Midwest Grain Products. Our Energy Partnership Proposal sought to assist Midwest Grain in improving their competitive position. It was a commitment to work with Midwest Grain as an "Energy Partner" to develop and implement a total energy service plan that would produce the most efficient utilization of resources.

The objective is to give Midwest Grain the option of enhancing the efficient use of energy resources, stabilizing their cost of production, and allowing them to concentrate their financial and human resources on their core business. This integrated package approach had numerous advantages for Midwest Grain.

- Long-term reliable electric and steam prices at attractive rates.

- An electric contract which limits exposure to the regulatory process while giving them the potential to benefit from gas-fired generation.

- An opportunity for Midwest Grain to manage their exposure to gas price fluctuations which is a major component in steam prices.
- Expanded energy services whereby CILCO would serve the customer's needs on both sides of the meter.

In order to respond to the customer's stated 1994 steam requirements, it was necessary to move expeditiously on the project. It was therefore intended that pending total project approval by the regulatory agencies, work and expenditures would commence utilizing CILCORP Inc. for financial resources and ownership.

The longer-range intent would be to transfer title of the project assets over to the utility company upon regulatory approval. CILCO, by taking advantage of the relationship with the holding company, would therefore respond in a timely manner to meet the energy needs of a key customer.

The regulatory process began immediately, but even so, it was too slow to permit CILCO to proceed alone in this transaction. It had always been intended to have ownership of the project within the utility. The question may be asked as to why we did not keep the project in the holding company. If the project was kept within a subsidiary of CILCORP, from a CILCO perspective, it would be yet another IPP. The subsidiary would face the same impediments/obstacles in our service territory that we would certainly present to other IPPs.

Current cogeneration technology makes the efficiency of producing kilowatt-hours extremely favorable. CILCO's System Planning Group has determined that cogeneration is the least-cost option for a facility on CILCO's system. This technology has been recognized as desirable by the Illinois Department of Energy and Natural Resources (IDENR) in a state-wide integrated resource planning process. The Illinois Commerce Commission (ICC) was very supportive of our approach and exhibited great flexibility and expediency in dealing with our petition.

# THE COGENERATION PLANT

CILCO, along with CILCORP Inc. (the utility's holding company), proposed to build a cogeneration plant on Midwest Grain's property consisting of three gas-fired boilers and one turbine generator. Two of the boilers would normally be operated with the third one on stand-by to provide 100 percent reliability for Midwest Grain's steam requirements. Each boiler would be capable of supplying 175,000 pounds of steam per hour.

The steam would pass through a 20 megawatt turbine generator being delivered to Midwest Grain. The plan configuration provides significant operating efficiencies and advantages. Instead of being designed to produce steam at 175 psi and 450°F, which is Midwest Grain's process steam requirements, the boilers are designed for 1,250 psi and 950°F. The incremental cost to do this is relatively low, since the boilers to produce steam for Midwest Grain are already needed. Heat rates of 4,800 Btu per kilowatt-hour are anticipated as compared to our normal system heat rate of 10,000 Btu per kilowatt-hour.

# FINANCING

In December, 1993, CILCORP Inc. announced that, through a new subsidiary, it agreed to finance the construction of the first phase of the $16.9 million cogeneration facility at Midwest Grain. The initial financing of the project by CILCORP Development Services Inc., a newly created subsidiary of the holding company, is for $11 million to construct steam boilers and other ancillary equipment.

This allowed us to provide for the steam requirements for Midwest Grain's $75 million expansion start-up in December 1994. CILCO sought Illinois Commerce Commission approval to invest in this project. Upon receiving the Commission's approval, CILCO acquired the steam boilers and other equipment from the subsidiary and invested an additional $5.8

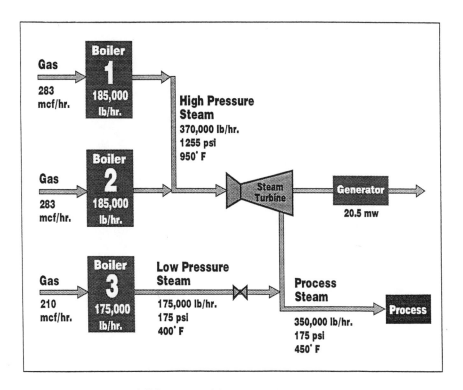

Exhibit 24-1. Midwest Grain Products

million in the project to install a 20 megawatt turbine generator that produces electricity through the cogeneration process earlier described. Commission approval was received in March, 1994.

The "energy partnership" is significant for both CILCO and Midwest Grain. It clearly demonstrates that we are capable of meeting the total energy needs of a major key customer in a non-traditional way. It clearly demonstrates that our organization is flexible and able to operate in a competitive marketplace. By providing this additional service and introducing a new product line, steam, to our current business, we will be able to provide savings to all of our electric customers. Start-up of steam service took place in December 1994; the electric plant went on-line in mid-1995.

*Chapter 25*

# New Prospects
# For Mini-Cogeneration

## Chapter 25

# New Prospects
# For Mini-Cogeneration

*Ronald E. Russell, Esq.*
Principal Consultant
Hagler Bailly Consulting, Inc.

*onservation* is the avoidance of spending an energy dollar. *Cogeneration*, on the other hand, is an effort to increase the velocity of
an energy dollar once the decision is made to spend one. For the
same dollar, you now get not only electricity, but also an additional
thermal energy benefit. Cogeneration involves the sequential production
of two forms of useful output energy from a single energy input. When
electricity or shaft power is the first output from the cogeneration system, the system is called a topping-cycle; when the thermal output is
produced first it is termed a bottoming-cycle.

A "bottoming cycle" requires a high temperature waste stream that
is relatively "clean" (i.e., not heavily contaminated with abrasive par-
ticles or other materials that can damage or foul heat exchange surfaces).
In terms of economic feasibility, the bottoming cycle typically 'competes'
against other uses for the waste heat (e.g., preheating combustion air or
producing hot water) that entail a smaller capital investment than would
cogeneration. Because of these constraints, bottoming-cycle systems are
seldom installed.

Topping-cycles are almost exclusively used in small-scale system
applications because they offer greater flexibility and utilize proven, low
cost, internal- combustion prime movers, such as reciprocating-piston
engines or combustion turbines. Large-scale cogeneration installations
are more likely to utilize external-combustion types of prime movers, but
all three types are used over a wide range of capacity ratings:

Reciprocating engines:     5 kW - 20 MW
Combustion turbines:       200 kW - 150 MW
Combined Cycle:            500 kW - 500 + MW

It should be noted that some of the steam produced from gas-turbine exhaust gases can be used to drive a steam turbine, and/or serve the host facility's thermal load.

# PACKAGED COGENERATION UNITS

The concept of "packaged" cogeneration units (PCUs), utilize compact internal-combustion engines rather than complex, custom-designed and field-erected steamboiler/turbine systems that are available. The entire system—engine, generator, pollution-control equipment, heat-recovery boiler and/or heat exchanger, and controls—are factory-assembled on a common structural-steel foundation, and completely tested before being shipped. The capital cost of a PCU installation per unit of output ($/kW) is lower than that of a custom-designed unit with the same capacity because it requires:

minimal site preparation
minimal space requirements
minimal construction time.

In addition, it uses standardized designs that enable economies of scale in manufacturing and tight quality control.

Although the installed cost of a PCU is less than that of a cogeneration unit using steam boiler/turbine system, PCUs entail the use of relatively high cost natural gas, propane or fuel oil as their fuel input. Steam boilers, on the other hand, can be fueled with lower-cost coal or waste materials—which may be free or even have negative cost (such as municipal solid wastes). There are many potential energy benefits of cogeneration such as steam, hot water, chilled water, cooling and depending on the type of fuel, society can benefit itself. Reducing municipal solid waste land fill, cleaning up contaminated water and soil, decreasing the density of a forest to a more optimum level and a plethora of other uses can result from cogeneration.

There are about 5 different technologies (prime movers) that could be used for self-generation: boiler/steam turbine, combined cycle, combustion turbine, reciprocating engines, and fuel cells. For known projects under development, combustion turbines and reciprocating engines by far account for the most commonly used technologies with combined cycle technology a distant third. In regard to capacity, a simple combustion turbine configuration is the most commonly used technology, with reciprocating engines next with half the capacity of combustion turbines and combined cycle capacity half of that of reciprocating engines.

The economics of these technologies will be impacted by the Clean Air Act (CAA), depending on geographical location and the type of treatment imposed on them in order to remain in CAA compliance. Federal and state air emission regulations recognize that reciprocating engines have an inherently higher level of $NO_x$ emissions than combustion turbines. Air emission regulations for reciprocating engines are also not as well-defined as they are for combustion turbines.

The disparity between the treatment of combustion turbines and reciprocating engines is driven by the technology-based approach to current air emission regulations in which a new facility is only required to install the appropriate "BACT" (best available control technology) within the particular type of "technology" being proposed. In other words, a new reciprocating engine must only meet the best emission level achieved by other reciprocating engines.

To date, only California requires an "alternative technology" analysis as part of its permitting process for new reciprocating engines. However, the possibility of more stringent emission controls (i.e., lowest achievable emission rate [LAER] technology does not allow any relief due to economic costs of technical feasibility) must be factored into the decision to use this type of prime mover.

Accordingly, cogeneration/self-generation may not be for everyone!

# EFFECTS OF COGENERATION SYSTEM CHARACTERISTICS

In addition to the three primary factors (electricity price, fuel price, and thermal load), the cogeneration system to be installed must have five important characteristics:

(1) The thermal and electrical outputs must be such that a significant fraction of the host facility's loads are satisfied.

(2) The total required investment must be "reasonable," considering all studies, designs, permitting activities, and all costs for equipment and on-site installation (including all equipment needed to supply and meet environmental regulations, to interface the thermal and electrical outputs and distribute energy produced to the end uses to be served, to provide supplementary electrical and thermal outputs at those times when facility loads exceed the respective output ratings of the cogeneration system, to provide full backup to the facility loads for uses at times when the cogeneration is not operational, and to monitor performance when it is operating).

(3) Operational efficiency should be high at high load factor (i.e., losses should be kept low, and a large fraction of the thermal energy output should be able to be used).

(4) Non-fuel operation and maintenance costs must not be excessive.

(5) Operational reliability should be high.

The first characteristic ensures that the PURPA criteria are satisfied (this presumes that PURPA-qualifying facility criteria [steam/electricity ratio] and the mandatory purchase requirement are not repealed) and that the electrical and thermal costs savings are as large as possible.

It must be noted, however, that there may be some special relationships between Items (1) and (2). First, large thermal and electrical loads mean that a relatively large cogeneration system can be installed. Economies-of-scale enable the cost per kW to be less than in the case of smaller systems, which tends to make the installation more economic.

Also, for some types of host facility, the most significant thermal loads are space cooling or refrigeration. In order for a cogen system to satisfy this type of host facility's thermal load, the system must incorporate an expensive thermal-powered chiller. (If such a chiller is already present, thermal load is generally included in the facility's steam load.)

Small cogen systems with a chiller are typically not economic, however, because (a) the first-cost of small thermal-powered chillers is high, and (b) the profile of the cooling load is typically rather peaked and its duration is limited to warm-weather months. Exceptions to the latter condition include facilities in which the cooling load is 500 tons or greater in all months, such as a large computer room, pharmaceutical plant, a frozen-food storage facility, etc..

A special point should be made concerning the relationship among fuel price and Items (2) and (3) above: If fuel price is low (e.g., coal, wood or waste material), then the total investment cost may be relatively high, and the penalty of a lower operational efficiency is not as great. Conversely, if fuel price is relatively high (e.g., natural gas or fuel oil), then the investment cost must be low and the efficiency high.

Similarly, there is an important relationship among Items (2), (4), and (5): In principle, every installation has an optimum level of reliability, but this is very difficult to assess given the individual components' probability of failure as well as the relative few projects upon which to base reliability studies.

Unfortunately, the consequences of not giving sufficient attention to ensuring that the design is adequate (i.e., reliability is at least 95% over an expected operational life of at least 10 to 20 years, with "reasonable" maintenance being performed) can be disastrous. Premature or catastrophic failure of a major part of the system may occur or costly preventative maintenance activities may be required. These activities are not only expensive as a direct cost, but also costly because of the downtime and replacement power required to accomplish the maintenance.

One aspect of the interplay between operational reliability and economics is illustrated by the question of the number of units to be installed in a given cogen system. Economies of scale generally mean that lowest first-cost is achieved by using a single unit, but the economic penalties associated with equipment failure can be large—particularly if demand charges are high and the tariff includes a demand ratchet. Multiple smaller units entail a higher capital cost, but maybe cheaper in the long run because a single failure will not lead to the loss of 100% of the system's output and the incursion of substantial demand penalties.

# FACTORS TO CONSIDER

The feasibility and attractiveness of a potential cogeneration installation are determined by many factors and considerations. These factors include regulatory, technical (including the magnitude, shape and temperature level of the thermal load), economic, and market considerations. The largest impetus to cogeneration over the last ten years has been regulations passed to remove obstacles to non-utility power generation, particularly the Public Utility Regulatory Policy Act (PURPA).

[CURRENTLY THERE ARE SIGNIFICANT DISCUSSIONS ONGO-ING REGARDING THE REPEAL OF SECTION 210 OF PURPA, THE MANDATORY PURCHASE REQUIREMENT.]

In addition, a variety of federal and state tax incentives (tax credits, exemptions, accelerated depreciation) and R&D programs were created that provided an important economic impetus to initial market development. Last, federal and state investment funds and state laws that protected the cogeneration project revenue stream were established to aid in financing projects, which increased the availability of capital for project development.

Utilities themselves have not been stagnant on the issues of non-host utility generation. They have taken significant action to reduce the motivation for self-generation. Utilities have become more adept in their ability to accurately compute avoided capital and energy costs; firm price contracts have been largely replaced with contracts that are linked to fuel prices; and more significantly they are studying the benefits of distributed generation.

Distributed generation theory suggests that the utility grid will have modular distribution webs, each module to be interconnected to adjacent webs with web networks interconnected to the preexisting regional grids. In other words, the electric utility industry will be transformed from large, vertically integrated systems into a series of smaller horizontally integrated units, each unit having the vertical aspects of production and delivery but also interconnected to other units of similar size and approximately located to load.

Another phenomenon that reduces motivation is that the market place has tended to focus on large cogeneration projects and independent power plants (IPPs), which has resulted in more power being offered to utilities than could be consumed. The resolution of this excess capacity has been a reliance on competitive bidding.

With regard to fuel price, in some localities a special natural gas tariff is available for cogeneration installations. Depending on the magnitude of the discount and whether the same low-cost gas can be used to fuel supplementary heating boilers and other equipment, these special tariffs can offer a strong incentive.

On the other hand, smaller installations tend to pay more than larger ones for fuel, and it often is not feasible to equip small units (under 700-1000 kW) with dual-fuel capability, and thus low-cost interruptible service rates cannot be used.

Impediments to self-generation:

• Power alone is difficult to evaluate—the need to value steam and other heat benefits.

• Small-scale projects are clearly dominated by systems that use natural gas as their primary fuel.

• Fossil fuel prices have declined and technological advances have increased efficiency and also reduced utilities' costs.

• State PUCs may allow different and/or negotiated rate structures.

In Michigan, the Detroit Edison Company, in response to the Big Three auto companies' effort to bypass the local distribution system, offered substantial negotiated rate decreases in return for assurances that no self-generation will be installed for the next ten years. The Michigan Commission accepted the tariff application which reduced electricity cost to the auto industry some $30-50 million overall.

The Commission deferred deciding who will pay for the reductions in future years. DetEd still has the opportunity to recover from other ratepayers if the then-sitting commission approves. It will have to show a cost of service study for recovery and also show that there was due discrimination. If it can't then shareholders will probably pick up a portion if not all of the discount. You can assume that the discount will match the projected savings from self-generation.

Other ways a PUC could frustrate the economics of self-generation is by designing declining block rates for certain load sizes and by disallowing inclusion of any self-generation in the computation of total load.

The PUC could also approve back-up power rates that can have the effect of making the self-generator pay twice for the same capacity—as part of the monthly capacity charge and again as part of the demand charge if the customer is forced to use utility power. This quasi-penalty will be premised on the customer/self-generator being responsible for assuming market risks and a policy discouraging the switching between market risk and franchise service protection (relying on the current concept of obligation to serve).

Implementation of cost based rates and/or reduction of buy back rates would also reduce the attractiveness of self-generation. Cost based rates, pegged to what the utility actually spends providing a particular type of service to a particular type of customer, would reduce the likelihood of bypass. The implementation of this practice may be closer than ever. It will be driven by the need to unbundle services for customers who want choice, the need for precision in pricing in a competitive environment, and the need for the types of information required to identify legitimate and verifiable stranded costs eligible for recovery.

# CONCLUSION

It is not clear whether self-generation is a feasible option for retail customers. With competition there may be less self-generation rather than more. As competition increases and the probability of deregulation of the generation sector increases, leading to supply capacity increases—electricity prices should decrease. These decreases will reduce the margin for gains from self-generation. Therefore the motivation for bypass will subside.

The National Association of Regulatory Utility Commissioners (NARUC) advocates doing away with PURPA section 210 only after full open access is available and state sanctioned competitive solicitation processes are in place. If the mandatory purchase requirement is repealed, the precision necessary to size a self-generation/cogeneration project becomes critical in order to realize the expected efficiencies.

If 210 is repealed, any excess electricity will have to be sold on the open market, probably at wholesale levels, provided the project can get an EPAct 211 access order from the FERC. The local utility will no longer be the purchaser of last or first resort.

It seems the question will be whether the expected savings from purchasing one's energy needs in the form of natural gas or some other fuel and generating their own electric power, thus bypassing the host utility's transmission and distribution embedded costs, outweighs the capital, fuel, and operation and maintenance costs for an entity who, generally, is not in the energy generation business.

*Chapter 26*

# Rule-of-Thumb Comparison of Efficiency: Small-Scale Cogeneration vs. Utility-Supplied Energy

*Chapter 26*

# Rule-of-Thumb Comparison of Efficiency: Small-Scale Cogeneration vs. Utility-Supplied Energy

*Bernard F. Kolanowski, President*
*Kolanowski & Associates*

T he essential advantages of cogeneration are often not understood by smaller commercial and industrial firms. This chapter is a simple explanation of the basic benefits which small-scale cogeneration can offer them.

In order to understand the benefits of cogeneration—the on site production of electricity and hot water—it is beneficial to know the overall efficiency of the energy media presently being used when compared to cogeneration.

Virtually every commercial and industrial establishment purchases their electricity from the local utility company and heat their water by using on site boilers and hot water heaters fired by natural gas or propane—which they also purchase from an outside supplier.

When on-site cogeneration is compared to purchased power the results in fuel usage efficiency are:

Cogeneration ........................... 89.2%

Purchased Power ................... 52.6%

When comparing this difference in efficiency to actual natural gas usage in a commercially available cogeneration system the following results are noted:

|                       | Cogeneration | Purchased Power |
|-----------------------|:------------:|:---------------:|
| Kilowatts             | 62           | 62              |
| Hot water, Btu's/Hr.  | 552,750      | 552,750         |
| Fuel Used, Btu's/Hr.  | 857,000      | 1,453,042       |
| Fuel Used Difference: | 596,042 Btu's/Hr. |            |

In a one-year period, with the facility operating 7800 hours/year, 4,648,000 cubic feet of natural gas can be saved, producing the same amount of usable energy.

Economically, the money that can be saved by cogenerating on site is:

|                       | Cogeneration | Purchased Power |
|-----------------------|:------------:|:---------------:|
| Cost of Electricity:  | $ -0-        | $36,248         |
| Cost of Natural Gas:  | $21,390      | $40,034         |
| Maintenance:          | $8,900       | $ -0-           |
| Total:                | 30,290       | $76,283         |
| Annual Savings:       | $45,993      |                 |

Another consideration to include in the benefits of cogeneration is that with less fuel burned, less products of combustion are released to the atmosphere, specifically carbon dioxide and nitrous oxide.

The overall result of on site, properly applied cogeneration is an economical, environmental, and conservational tool that preserves an establishment's cash, helps reduce pollution and conserves a precious natural resource.

# APPENDIX

The following parameters were used in compiling the above:

Cogenerator Specifications: Intelligent Solutions, Inc. Model ISI601. 62 kilowatts; 552,750 Btu's/Hr. hot water output; 857,000 Btu's/Hr. fuel input.

| | |
|---|---|
| Hot Water Heater Specifications: | 70% thermal efficiency |
| Btu's/Kilowatt: | 3,415 |
| Electrical Production: | 10,700 Btu's/Kilowatt |
| Btu's/Cubic Foot of Natural Gas: | 1000 |
| Btu's/Therm: | 100,000 |
| Hours/Yr. of Cogenerator Operation: | 7800 |

*Chapter 27*

# Success Stories about "Inside-the-Fence" Cogeneration

*Chapter 27*

# Success Stories about "Inside-the-Fence" Cogeneration

*Martin Lensink, P.Eng.*
*Manager, Canadian Operations*
*U.S. Turbine Corporation*

I nside-the-fence" cogeneration is a younger industry in Canada than in the U.S. The four main success stories which are reported in this article refer to projects developed when I was with the engineering consulting firm of SNC/W.P. London, and most recently with U.S. Turbine.

Inside-the-fence cogeneration is defined here as systems which are owned and operated by the steam and power user, with little, if any, surplus power sold to the electric utility or wheeled through it.

At U.S. Turbine, we define a successful "Inside-The-Fence" project as one that:

• has compatible electrical and thermal loads over the entire operating regime;

• was brought on line on schedule and within budget;

• has an efficient thermal cycle, with no or little recoverable heat being wasted;

• is reliable, with forced outages minimized and utility outages isolated;

- is maintainable, i.e., the frequency, cost and duration of system downtime have also been minimized;

- is flexible, in terms of starting, stopping, and restarting quickly;

- is simple and easy for operators to monitor and control; and

- does an automatic derivation on a daily or weekly basis of net energy cost savings.

# M.H. HEINZ COMPANY OF CANADA, LIMITED

The first inside-the-fence cogeneration success story highlighted here is the Heinz Factory in Leamington, Ontario, about 3/4 of an hour into Canada from Detroit. (Figure 27-1.)

This cogeneration system consists of two combustion turbine gen sets, Allison Model 501-KB5 rated nominally at 3.7 MW each, with a maximum turbine inlet temperature of 1895°F. Behind each turbine gen set is a waste heat boiler with duct burner producing 120 psig steam. Each waste heat boiler raises about 24,000 lbs/hr in the unfired mode, and up to 64,000 lbs/hr in the fired mode.

The project consisted of removing three old air wall type boilers, and replacing them with the two waste heat boilers discussed and adding the two gen sets. The gas turbine gen sets are located outside the Heinz boilerhouse with a roof extension to support the turbine inlet air filters and lube oil coolers.

The two D type HRSGs are shown here. One key factor contributing to the success of this project, was that the Heinz Project Manager/ Project Engineer was the Powerhouse Chief and Manager of Utilities. His knowledge of powerhouse systems, combined with his utility experience, was a valuable asset to the project. The project was developed during 1987/88 with engineering and construction taking place a 10-month period in 1989/90. The system was fully commissioned in August of 1990.

Another factor contributing to the success of this installation was the 225 psig natural gas line constructed by the LDC, eliminating the need for an inlet fuel gas compressor. Heinz considers the installation to be successful because it was built on budget and on schedule, with pre-

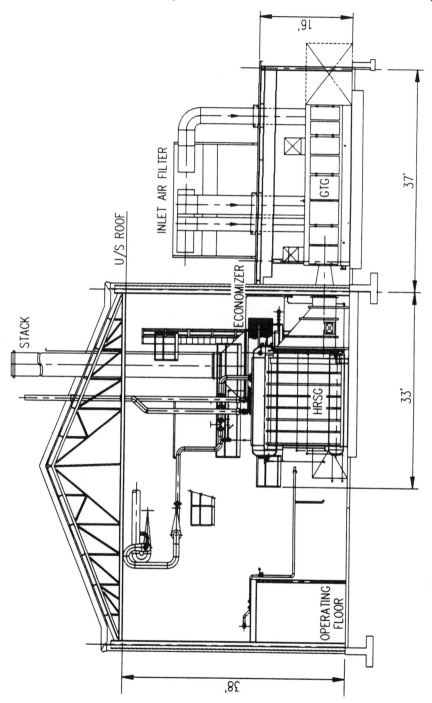

**Figure 27-1. H.J. Heinz Company of Canada, Limited (Leamington, Ontario)**

dictable annual energy savings. Maintenance costs have been lower than industry "rules of thumb."

# UNIVERSITY OF WINDSOR

The University of Windsor looked at inside-the-fence cogen for about six years before deciding to proceed.

Their system consists of one turbine gen-set, Allison Model 501-KB5, nominally rated at 3.7 MW; an unfired HRSG with a capacity of 25,000 lbs/hr; and two absorption chillers rated at 800 tons each. Incorporating the existing single-effect absorption chillers, which were previously "Last On - First Off," and which are now "First On - Last Off," was a key facet of this project's success. The existing boilers raise the additional steam needed between November and March.

The major equipment was located inside the existing powerhouse. When the powerhouse was designed and built in 1972, this space was intended for a steam turbine-generator. Use of existing real estate contributed to the capital cost success of this project. (Figure 27-2.)

The HRSG provides the base load demand of the Campus District heating and cooling system. All the cogenerated electricity is used on campus. Again, the LDC supplied 225 psig gas.

# LABATT BREWERIES OF CANADA

Labatt Breweries began to look at an inside-the-fence cogen system for their London Plant in 1989.

The system, which was commissioned late in the Fall of 1993, produces 4.9 MW nominally from an Allison Model 501-KB7 and up to 100,000 lbs/hr of 115 psig steam from a single O-type HRSG. The HRSG was sized to meet the future peak demand of the brewery.

The gen-set was sized to meet the plant electrical demand on a hot summer day (4.2 MW). As a result of this design criteria, there is surplus power, which is sold to the electric utility. (Figure 27-3.)

Since Labatt's wanted the major equipment to be installed on the same elevation as their existing boilers and refrigeration equipment, the unit installation cost (i.e. $/kW) was somewhat higher than the $1000/kW "rule-of-thumb," specifically:

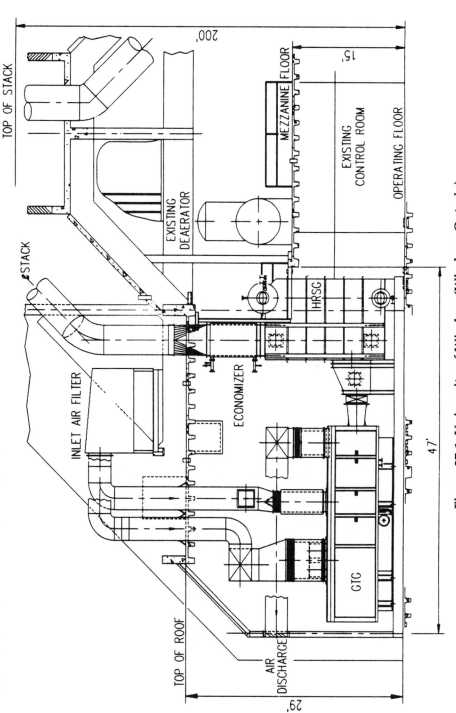

**Figure 27-1. University of Windsor (Windsor, Ontario)**

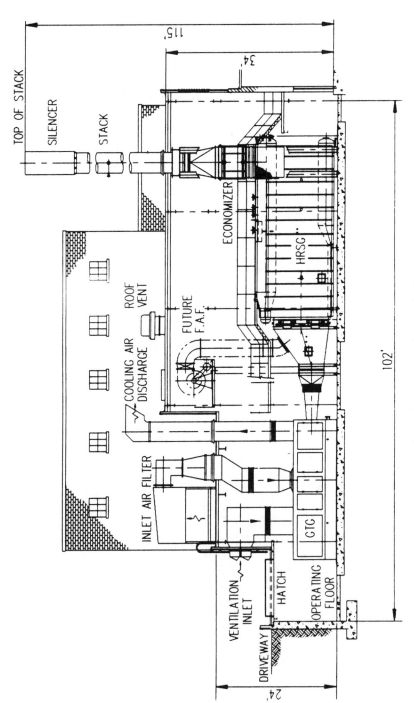

**Figure 27-3. Labatt Breweries of Canada (London, Ontario)**

• Site preparation costs including relocation of buried services, excavation.

• Cost of retaining walls was significant.

• Although the local natural gas utility constructed a 225 psig gas line for the combustion turbine, available pressure was not quite high enough. A small fuel gas compressor to boost the turbine supply pressure about 50 psig was still required.

# SONOCO LIMITED

The final on-site cogeneration system is a paper mill installation which we at U.S. Turbine completed two years ago. (Figure 27-4.) This system consists of an Allison 501-KB5S gen-set nominally rated at 4 MW, and a single O-type HRSG with capacity of 55,000 lbs/hr at 115 psig. The LDC supplied 175 psig gas. Again, much of the customer's balance-of-plant steam and electrical equipment was retained in the cycle, keeping the installed cost to a reasonable level.

Although there were some unexpected site preparation and civil costs, they were not nearly as significant as the previous installation.

The HRSG was erected first, and the GTG immediately thereafter. Total erection time for the major equipment was about 1 week. The structural steel was added around and over the major equipment.

The system was sized on present base load steam demand and predicted future peak. This results in some surplus electricity, which is wheeled to a sister paper mill in Ontario.

This system was constructed on a single contract design build turn-key basis in about 10 months between March and December of 1993. Although it had logged only 3,000 hrs (at the time this article was written), it meets all facets of our success definition. Even our Crown Electric Utility, Ontario Hydro, considers it to be successful.

U.S. Turbine had one engineer on site during the entire construction period. He provided liaison between the home office engineers and the subcontractors.

Construction costs were on budget. About 190 carefully thought out construction drawings were prepared. When the trades had interpre-

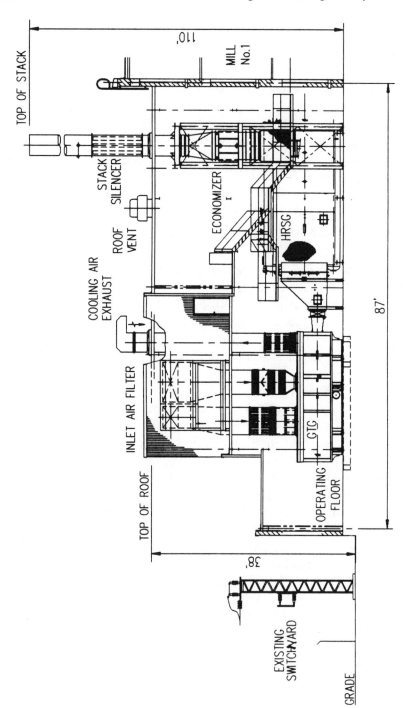

**Figure 27-4. Sonoco Limited (Branford, Ontario)**

tation questions, U.S. Turbine's site engineer could clarify them quickly.

---

Based on my 8 years in the "Inside-The-Fence" cogen business, here are some observations I would like to share with you:

- Is the project driven internally and managed by the Powerhouse Chief or a Staff Engineer?

- Construction strategies very widely between totally traditional and total turnkey with several variations in between.

- Somewhat related to the first point, is that important objectives, such as maximizing uptime and lowest calculated life cycle and first costs can yield different results.

- Some customers view "inside-the-fence" cogen as being like any other capital project. If you have "done" a few of these, you will quickly conclude that this is <u>not</u> the case.
- Customers vary widely between, on the one hand, a corporate culture that keeps suppliers and contractors at arm's length, versus on the other hand, a corporate culture that promotes establishing strategic alliances.

- We have also found over the years that even though the GTG supplier most often supplies only a gen-set, he ends up commissioning the whole cogen system.

- "Inside-The-Fence" cogen usually must do more than save money and have a payback. It must solve a bigger problem or issue.

**In conclusion, critical success factors for "Inside-The-Fence" cogen include:**

- A knowledgeable proponent, who has seen a lot of installations and has talked to a lot of people.

- A knowledgeable implementation team, and I would now go so far as to say, have the same individuals accountable for all phases of the project, including design, procurement, construction, <u>and</u> start-up to prevent change orders and delays, and provide areas of responsibility and immediate direct feedback to design.

- A corporate will to "do" cogen even though it is not their main line of work.

- Developing a good commercial and technical relationship with the electric utility. A clean site.

- Doing a lot of homework on how to minimize turbine downtime.

- Deliberate and diplomatic negotiations with the electric utility to establish fair standby power costs.

**My recommendations to potential "inside-the-fence" cogenerators are:**

- Compare your internal capital project policies; corporate culture; and resources to the critical success factors listed above.

- Be convinced that a quality system with predictable energy savings and very high availability over the long-term is achievable.

- After seeing as many installations as possible and talking to the people who run them, develop a strategy to manage for success.

- One component of this strategy will be to compare one strategy which is the lowest cost at each project milestone, such as consulting services, major equipment, minor equipment, and construction trade agreements versus a purchasing strategy that adds and creates high value.

- The fundamental decision with respect to successful "inside-the-fence" cogen is what are we really buying: A gen-set, then a boiler, then construction services?
  Or are we buying a fully-functioning system with risks borne by the people who are in the best position to manage those risks?

*Chapter 28*

# Case Study: Meeting Changing Conditions at the Rhode Island Medical Center Cogeneration Plant

*Chapter 28*

# Case Study: Meeting Changing Conditions at the Rhode Island Medical Center Cogeneration Plant

*Donald P. Galamaga, C.C.P. and Paul T. Bowen, P.E.*
*Rhode Island Department of Mental Health,*
*Retardation & Hospitals*

## OVERVIEW—THE DEPARTMENT AND THE CENTRAL POWER PLANT— A SPECIAL PARTNERSHIP

In 1902 the Central Power Plant (CPP) was established at the Rhode Island Medical Center (RIMC) to provide heat and electricity for the facility and to make the Center self-sufficient. The plant consisted of coal fired boilers and steam engine driven generators.

In 1936 a new cogeneration plant was built and consisted of two General Electric Company steam turbine generators rated at 750 kW - 0.80 PF. Throttle conditions were 400 psig and 600 degrees F with steam extraction at 100 psig and exhaust (back pressure) at 15 psig. This plant has been in continuous operation since 1936.

A 1956 expansion added two Elliott Company steam turbine generators rated at 2,000 kW - 0.80 PF and a 120,000 #/Hr. Riley Corporation steam boiler.

The Central Power Plant operated as an isolated plant until 1974, at which time it was interconnected with the Narragansett Electric Company (NECO) system. The interconnection was required due to increased imbalance of the steam and electrical loads.

The facility served by the CPP is a 200-acre site off Route 95, three miles south of Providence. There are 160 buildings and structures of various types on the site. The buildings house programs of the Departments of Administration (DOA), Corrections (DOC), Children, Youth & Families (DCYF), Mental Health, Retardation & Hospitals (MHRH), and the Rhode Island Lottery. Presently there are 3,000 inmates, patients, and over 6,000 staff and workers at the site. Total building floor area is approximately 2,800,000 square feet. See Figure 28-1.

# FUEL PURCHASING

The fuel usage at the RIMC is shown in equivalent barrels of #6 oil; Figure 28-2 shows usage in recent years. The high usage in the late 1960's

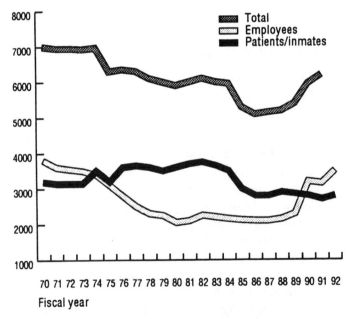

Figure 28-1. Population at RIMC

and early 1970's is due to a combination of deteriorated distribution systems (leaks), the steam dumped to the atmosphere to meet electrical loads prior to the 1974 connection to NEC, and the inefficiency of older 1933 boilers.

Oil consumption estimates prepared in February 1992 project the fuel usage will increase to 190,000 equivalent barrels for the 1993-1994 season, see Figure 28-2.

The boilers at the CPP are equipped with combination gas/oil burners, except for Boiler #6, which is equipped only for oil. Oil used is #6,1 percent sulphur. Natural gas is generally supplied by the Providence Gas Company on an interruptible basis.

Past practice was to bid #6 oil on an annual basis. Pricing was based on the lower (day of delivery) of the Boston or Providence posted Tank Car Schedule as published in the New York Journal of Commerce, plus or minus a fixed amount determined by the bidders. There is no fixed quantity of purchase, though bid data show historical usage.

The first serious attempt to purchase natural gas competitively occurred in the summer of 1987 as a result of discussions with gas suppli-

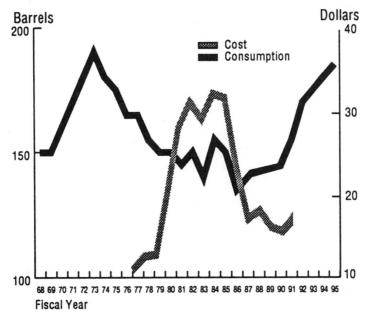

**Figure 28-2. RIMC Central Power Plant Fossil Fuel Consumption (Thousands of Barrels) and Fuel Cost (Dollars Per Barrel)**

ers. The first contract was initiated on a monthly basis with the Providence Gas Company in September, 1987. There exists a problem with the rates approved by the Public Utilities Commission (PUC), which permits a variable charge for local transport from $.10 to $.85 per thousand cubic feet of gas. Since September, 1987 MHRH has purchased gas from various suppliers with payment to the LDC for local transport by the supplier or the State. While MHRH is certainly not completely satisfied, due to variable transport rates charged by the LCD, it has forced the price down so that generally gas is the fuel used from April through October. MHRH has burned natural gas as late as December 15 in some years.

People who have worked for or with governmental agencies will appreciate the effort required by the people in the Departments of MHRH, Purchasing and Administration to overcome historical practices and required public bidding procedures.

Increased gas pipeline capacity and new distribution systems within Rhode Island should improve competition and result in lower gas prices.

The availability of pipeline natural gas has resulted in increased competition between oil and gas, as well as competition between the local gas company and gas supply companies. All of this has been beneficial to our fuel budgets. Figure 28-2 shows the annual cost of fuels per equivalent barrel of #6.

# STEAM ABSORPTION CHILLER

Because of the invested capital in the steam distribution system and our lower summer steam loads, the incremental cost for use of steam absorption chillers for air conditioning is minimal and improves the cogeneration. Figure 28-3 is a typical analysis of the comparative economics for a 200 ton electric and steam absorption chiller. It will be noted the added capital investment for the steam absorption chiller has a simple payback of 4.0 years. The electrical demand reduction is an important element in the analysis due to impact on the building electric system design, secondary transformer size, distribution system capacity and the CPP switchgear capacity. There are a total of 2600 tons of steam absorption units installed at the RIMC, ranging in size from 100 to 448 tons.

---

**1.  Capital Investment**

Absorption Chiller, 200 ton installed ................................................$160,000
Electric Chiller, 200 ton installed ..................................................102,000
Capital Cost Difference................................................................$ 58,000
Reduction Transf. Size, 150 kVa ..................................................$ 4,000
Net Difference ...........................................................................$ 54,000

**2. Operating Cost**                                                       **Cost/Yr.**

A.  Electric Chiller

   Electricity - (200)(1500 FLH)(0.10) = ..................................$30,000
   Electricity Demand - (200)(2)(11) = .....................................$7,200
                                                                          $ 37,200

B.  Steam Absorption

   $$\text{Steam} - \frac{(200)(18)(1500) \times 5.00}{1000}$$

   ...................................................... $ 27,000

   Cogenerated Electricity (Credit)

   $$\frac{(200)(18)(1500) \times 0.45}{30}$$

   ....................................................$8,000

   Net Operating Cost .................................................................$19,000
   Added Debt Services (Absorption Unit)..................................$4,700
   Operating Cost .......................................................................$23,700

**3. Simple Payback**

   Operating Cost Difference .......................................................$13,500
   Capital Cost Difference............................................................$54,000
   Simple Payback =  58,000 = ................................................4.0 years
                      13,500

---

**Figure 28-3. Comparative Costs 200-Ton Electric and Absorption Chiller**

# CPP STEAM CYCLE &
# STEAM DISTRIBUTION SYSTEM

The steam distribution system for the RIMC consists of over 20 miles of steam and return piping. Steam is distributed at two pressure levels, namely 100 psig and 20 psig nominal. The higher pressure steam was originally dictated for sterilization, cooking and laundry requirements for the hospital. In recent years changes in operation have resulted in less need for high pressure steam.

The CPP Flow Diagram is shown in Figure 28-4; steam is extracted from the turbines at 100 psig and exhausted at 20 psig into the distribution systems. An analysis of the Elliott steam turbine generator performance data indicates as a simple rule that 60 pounds of steam are required per kWh when extracted at 100 psig, and 30 pounds are required per kWh when exhausted at 20 psig. These figures are higher than the original performance data on the Elliott turbines and allow for lower performance due to age. Stating the operation in another way, twice as much power can be cogenerated with 20 psig steam as with 100 psig extracted steam.

Two areas where changes have been made to improve cogenerated power follow. First, the laundry, constructed in 1958, had a criteria for 150 psig steam, which could only be obtained by reducing of 400 psig steam and without any cogeneration. Review and tests revealed the laundry could operate satisfactorily with 100 psig steam system. The piping system was modified to utilize 100 psig steam, which provides cogeneration power of 640,000 kWh per year, a saving of $50,000 with a simple payback of 0.5 years.

Secondly, the General Hospital Regan Building has a 250 ton steam absorption chiller, which had been run on the 100 psig steam system since start-up in 1981. In 1989 an investigation was made to determine reasons for inability to operate on low pressure steam. It appears the problem was due to lack of trapping on the unit and the arrangement and trapping on the steam piping.

Modifications to piping and trapping resulted in satisfactory operation on low pressure steam, providing an additional cogenerated power of 107,000 kWh per year, a savings of $4,800, and a simple payback of one-half year. Other areas are being investigated.

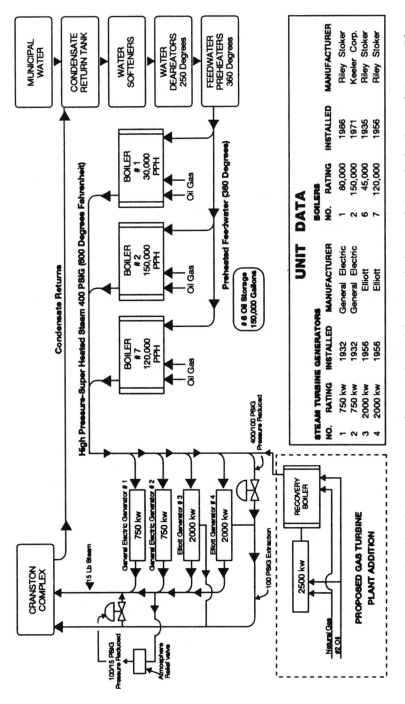

Figure 28-4. Rhode Island Department of Mental Health, Retardation and Hospitals Central Power Plant, Cranston, RI

# UTILITY COMPANY COOPERATIVE AGREEMENT

In 1990 the Department of MHRH entered into a Cooperative Interruptible Agreement with the Narragansett Electric Company. Under the terms of this agreement, MHRH would, "on call" from the utility, reduce its purchased power demand by 500 kWh in the summer and 300 kWh in the winter. This agreement resulted in a reduction of $100,000 in electrical charges for the year or about $.00625/kWh of purchased electricity.

To meet load reduction requirements, MHRH had three alternatives—reduce load by use of building emergency generators, load shedding, or increase steam turbine generation by exhausting low pressure steam to the atmosphere. The latter alternative is four times more expensive and least desirable.

# 1936 STEAM TURBINE GENERATOR UPGRADE

The General Electric Company steam turbine generators generally were little used in recent years until about 1988 when problems with the Elliott Turbines resulted in loss of power during the winter months.

The turbines have been operated at reduced loads, 300 kW. In 1990, as a result of a major short circuit, the bus bars, switches, instrument transformers and associated items were damaged.

In connection with repair work, bids were taken, and General Electric was awarded a contract to test and repair the generators, exciters and switchgear and to provide new controls and instrumentation for synchronizing the generators. The work was completed in 1992. Tests indicate the generators are capable of 800 kW, 0.90 PF. The cost of reconditioning for both units was $130,000. Upgrade of the turbine generators has enhanced their reliability and increased the total generating capacity by 900 kW, a cost of $144 per kW.

# PLANT EXPANSION ALTERNATIVES

A study of the RIMC steam loads and total electric load requirements indicates the electric load is increasing faster than the steam load. This results in less opportunity for steam cogenerated power and increased quantities of the more expensive purchased electricity. See Figure 28-5. Plant expansion alternatives could include diesel engine generators or gas turbine, both with heat recovery. A gas turbine with heat recovery at steam conditions to permit use in the existing steam turbines (a combined cycle) would appear most attractive; this cycle is indicated in Figure 28-4.

A simplified feasibility analysis prepared in 1988 (Figure 28-6) indicates a simple payback of 3.7 years based on a single 2500 <W gas turbine generator. Electric usage has increased more than anticipated, which should improve the economics for gas turbine generation.

In-depth studies must be made for final selection of new cogenerating unit(s), which would address the many issues involved.

# PRIVATIZATION STUDIES

While the evaluation of plant expansion alternatives has been proceeding, the state has also engaged the Army Corps of Engineers on a project which is seen to be both mutually beneficial to the engineers and

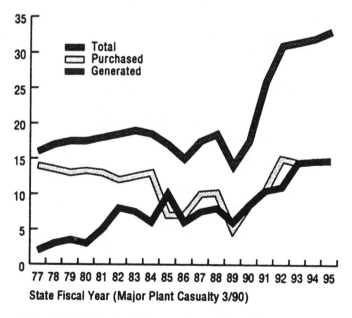

Figure 28-5. RIMC Central Power Plant Electricity Generated & Purchased

COGENERATION PARAMETERS

1.  ASSUMED ELECTRIC & THERMAL LOADS
    Electric Usage (1993) = ..........................................28,000,000 kWh
    Steam Usage (1993) = ...........................................850 MM Lbs.

2.  STEAM/ELECTRIC RATIO # Steam/kWh
    Lbs. Steam/kWh =   850 × 10 MM =...................30.35
                       28 × 10 MM

3.  STEAM GENERATION PARAMETERS
    Steam Turbine Generation ....................................55 lbs./kWh
    (Assume Extr. Operation)
    Gas Turbine Generation.........................................5 lbs./kWh

4.  ELECTRIC/STEAM BALANCE
    The combination of steam and gas turbine electric generation must
    result in 30.4 lbs. of steam\kWh to satisfy the thermal load.
    If "X" equals the kWh of Steam Turbine Generation and an adjust-
    ment factor (5/55 =.09) is made for steam turbine power from the
    HRSG (Gas Turbine), then:
    $$(X)(55) + (1-X)(5)(1.09) = 30.4$$
    $$X = ....................504$$
    i.e., 51% of the electric generation should be from the steam tur-
    bines and 49% from the gas turbine
    Turbine Electric Generation = ........................................14.3×10MMkWh
    Turbine Steam Required = 14.3 × 10MM × 55 = ........786.5 MM Lbs
    Gas Turbine Electric Generation = ...............................13.7×10MMkWh
    Gas Turbine Steam (HRSG) = 3.7 × 10MM × 5 = ......68.5 MM Lbs.

5.  COST OF STEAM & POWER
    **Case 1- Present System**
    Base Fuel Cost: Oil #6 1% = ...............................$18.00Bbl.

    $$Cost/M\,Lbs. = \frac{(1,000)(1,028)(18.00)}{(150,000)(42)(.82)} =$$
    ....................$ 3.58

    Purchased Electricity=..........................................7.5 Cents/kWh

    $$Fuel\,Oil\,\#2\;\frac{(1,000,000)\times.54}{(138,500)\quad.95} =$$
    ......................$4.10/MMBtu (LHV)

    Boiler Fuel = (850)(1,000)(3.58) = .........................$3,043,000

    $$Elec.\,Gen. = \frac{(850)\,(1,000,000)}{(55)} =$$
    ............................5.455MMkWh

Elec. Purchased = 28.0 -15.455 = .........................12.545 MMkWh
Elec. Cost = (12,545)(1000000)(.075) = .................$940,875
Total Energy Cost = ...............................................$3,983,875

**Case 2- Gas Turbine (95% Availability)**

1. Gas Turbine Fuel
   Elec. Gen. = ...................................................13,600,000 kWh

   $$\text{Avg. Load} = \frac{13,700,000}{8,420\,\text{Hrs.}}$$
   ...................................1627kW

   Gas Turbine Fuel Rate = ...............................21.7 MMBtu/Hr.
   Fuel Usage = 8,420 Hrs. × 21.4 = .................182,714 MMBtu
   Fuel Cost = (182,714) (4.10) = .......................$749,128

2. Steam Turbine Fuel Cost
   Gen = (14.3) (100,000) (55) = .........................786.5
   Fuel Cost = 786.5 × 1000 × 3.58 = ................$2,816,000

3. Total Energy Cost
   Steam Generation.................................................$2,816,000
   Gas Turbine.........................................................522,800
   Total $3,338,800
   Credit Electric Cost Saving ..................................$- 878,150
   Total.................................................................$2,460,650
   O&M Expense (Exc. Fuel)......................................-137,000
   Net Cost................................................................$2,597,650

4. Savings - Case 2 vs Case 1
   1.Case 1 ........................................................$3,983,875/Yr.
   Case 2 .................................................................2,597,650
   Savings.............................................................$1,323,500/Yr.
   2.Simple Payback
   Gas Turbine Capital Cost
   (2500 kW) (1800) = ...........................................$4,500,000
   Annual Savings.................................................$1,323,500

   Simple Payback
   4,500,000 ..........................................................= 3.7 Yrs.
   21,323,500

**Figure 28-6. Rhode Island Medical Center Central Power Plant Gas Turbine Cogeneration Analysis**

to the Department of MHRH. The engineers are looking to develop further expertise in the area of utilities analysis and cogeneration and the state is looking for some confirmation or other options regarding who should run the Central Power Plant in the future.

To this end, a study was commissioned where the state and the Army Corps of Engineers share fifty-fifty in the costs of an overall project aimed at determining or validating the future requirements for the Central Power Plant at the Rhode Island Medical Center. This study was supplemented with a cost benefit analysis to indicate whether the operation should be taken over by the private sector or whether there are more savings accruing to the state if it continues to run the project itself.

# CONCLUSION

In conclusion, it is quite evident that the Rhode Island Medical Center Central Power Plant and its utilities distribution system are a major resource to the State of Rhode Island and a major benefit to the taxpayers of the state. The direct beneficiaries of the savings realized by the Central Power Plant have been Rhode Island's most disabled citizens. Resources saved by the Rhode Island Central Power Plant have been allowed to be applied to programs for the chronically ill, the severely mentally ill, and the mentally retarded in the state with the result that leadership in the overall Department and specific leadership within divisions has been able to focus on program development and program services for these patients and clients. The result has been a set of nationally recognized programs with an unusual cogenerating partner.

*Chapter 29*

# Case Study:
# A Public-Private Partnership
# In Cogeneration

*Chapter 29*

# Case Study:
# A Public-Private Partnership
# In Cogeneration

*Rita Norton, Nayeem Sheikh, John Reader*
*City of San Jose, California*

The City of San Jose's Office of Environmental Management co-ordinated the installation of a 1.5 megawatt cogeneration sys-tem in its new convention center for approximately $2.5 mil-lion. The convention center cogeneration system represents the culmi-nation of a six year effort to forge a partnership between the public and private sector to meet the future energy needs of both parties efficiently and economically.

In this innovative public-private partnership, the city has devel-oped an agreement with its private sector neighbor, a major hotel com-plex, to sell the excess thermal energy generated by the cogeneration plant. The city purchases natural gas at a discounted cogeneration rate from Pacific Gas and Electric Company (PG&E) or from open gas market.

The convention center central plant cogeneration system, a state-of-the-art, environmentally sound system, produces heat and chilled water for the convention center and two adjacent facilities, which together oc-cupy over 1.5 million square feet. In addition, the system supplies elec-tricity to the city's convention center and the adjacent 100,000-square-foot main library; the remaining energy is sold to PG&E.

The convention center cogeneration program has allowed the city to demonstrate the success of a public-private partnership in cogeneration. As a result, the system is being used as a marketing tool to attract private development in the city's downtown district.

Pleased with the results of this program, the city is in the process of developing guidelines to promote the use of cogeneration systems in new and renovated buildings in the city. These guidelines will be available free of charge to other public agencies and the private development community.

# DIRECT IMPACT OF THE PROGRAM

This innovative program has energy, environmental and economic features. The convention center central plant cogeneration system is a state-of-the-art and environmentally friendly system which is meeting the energy requirements of three adjacent facilities, which occupy about 1.5 million square feet.

The cogeneration system supplies both electricity and heat and chilled water for air conditioning to the San Jose Convention Center and Main Library and heat and chilled water to the Hilton Hotel.

The system reduces the cost of operation of the facilities it serves. It also increases the hotel's marketability. The convention center and the library are saving over $500,000 a year in utility costs. These cost savings will increase as energy costs increase.

The city is realizing a 3.2-year payback on its initial incremental investment in the 1.5-megawatt facility. The total cost of central plant was $2.6 million (in 1986 dollars) of which $.9 million was used to purchase conventional boilers and chillers. An additional $1.6 million was used to purchase and install the engine generator and absorption chiller.

The city can use this case study as an example to demonstrate that cogeneration systems:

1. are cost effective,

2. are capable of meeting exacting air quality standards,

3. can provide significant ancillary side benefits such as solid waste reduction and resource recovery.

# SIGNIFICANT IMPACTS ON THE QUALITY OF LIFE OF THE CLIENTS WITHIN SAN JOSE'S JURISDICTION

The convention center central plant cogeneration system has energy, environmental and economic impacts.

The System meets the energy requirements of three adjacent facilities, namely the San Jose Convention Center, the San Jose Main Library and the Hilton Hotel complex. Environmentally, the cogeneration system meets exacting air quality standards and provides significant ancillary side benefits such as solid waste reduction and resource recovery. The system has proved to be economically beneficial to both the city and its private sector partner, demonstrating the feasibility of a successful public-private partnership in energy utilization.

The clients who are benefiting from this program are the city's citizens private businesses, and the environment.

(1)   Since the city is saving more than $500,000 a year in avoided utility costs, the system essentially "generates" an equivalent amount which may be used to either reduce the general tax burden or to fund other city programs.

(2)   The Hilton Hotel is benefiting from this cogeneration system. The thermal energy required for air conditioning and heating of this hotel is supplied by cogeneration system at a discounted rate which is cheaper than using PG&E's power supply. Also the hotel owners avoided about $400,000 in capital cost by not having to buy major air conditioning and heating equipment. The hotel owners also saved space required to house this equipment.

One of the most innovative and effective aspects of this program is the public-private partnership which combines the energy needs of several municipal buildings and the adjacent private sector business. This system has realized the following overall benefits:

• Increased energy efficiency and renewable energy use.
• Reduced costs of operation and maintenance for both the public and private sector.

- Enhanced local control over energy supplies and prices.
- Improved air quality for the community.

In summary, this program is effective because:

(1) The actual performance has exceeded predicted performance.

(2) The system uses one input and gives three outputs. The input is natural gas, and the outputs are electricity, hot water and chilled water. Conventional power plants are about one half as efficient, as shown in Figure 29-1. (This also translates into the fact that the cogeneration systems can be 50% more friendly to the environment.)

(3) It provides benefits to both the public and private sectors.

(4) In technical terms, the program system uses 165 therms of natural gas every hour and produces out 1500 kW of power and 6,000,000 Btus of heat every hour.

(5) In financial terms, on an hourly basis, the system costs $48 an hour and produces $190 of energy every hour as shown in Figure 29-2. ($150 of electricity and $40.00 of heat energy).

(6) This system produces electric energy at an average cost $0.04/kWh compared to a direct purchase at $0.10/kWh from PG&E.

## OPERATIONAL CHARACTERISTICS

The convention center maintenance staff and building users are satisfied by the quality of utility operation. Building users are provided with lighting, heat, air conditioning, hot water and so forth as they would in any modern building. They are unaware of the system which is as it should be. The convention center's departmental utility budget is reduced considerably. The technology used is fairly common, so the in-house maintenance staff can maintain the system without the help of expensive outside maintenance contracts.

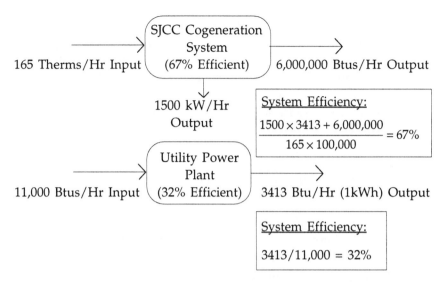

Figure 29-1. Efficiency Comparison of Power Production Methods

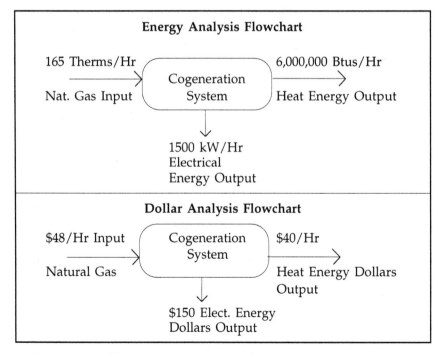

Figure 29-2. Efficiency Comparison of Power Production Methods

The fact that the Hilton Hotel is using energy from the cogeneration is an evidence of the support for the program. There are savings to the hotel owners in initial capital equipment cost space savings and the daily operational cost (utility and equipment maintenance). The adjacent library building was paying $260,000 per year for its gas and electric expenses before the cogeneration system was installed. The project has reduced this expense to $108,000 a year.

## TRANSFER OPPORTUNITIES

Public-private partnership in cogeneration can be accomplished in other regions where both parties have significant energy requirements and are located close to each other. Typical payback periods are between two to four years depending on the level of energy used. The life of the project itself is about 40 years.

Easily performed studies can determine whether the use of a cogeneration system is economically feasible. In most applications where the use of hot water is common, a cogeneration system can be a lucrative option. Any project which uses hot water, air conditioning and electricity can consider a cogeneration system to reduce the operational costs. If replicated by cities across the nation, total energy required to sustain future needs would be reduced. Such investments would be beneficial for local governments, private businesses and future generations.

## PROGRAM IS INNOVATIVE

Cogeneration itself is not a new concept in the private sector. However, for municipalities, who have traditionally relied on utility companies, cogeneration is relatively new. The additional innovative aspect of the City of San Jose's cogeneration efforts is the identification of a downtown site which allows the benefits of cogeneration to be shared by a municipality and its private sector neighbor.

Through advance planning, the city identified opportunities for a public-private partnership which resulted in monetary benefits to the individual parties as well as in the protection of the environment.

# OVERCOMING ADMINISTRATIVE, PROCEDURAL AND LEGAL OBSTACLES IN IMPLEMENTING THIS PROGRAM

The major obstacles in implementing the public-private partnership in cogeneration included:

• Obtaining the support of City Council.

• Coordinating of facility planning.

• Negotiating with the private business.

• Securing the necessary permits and energy agreements with the utility, and the Bay Area Air Quality Management District.

• Financing.

These obstacles, and those relating to reliability concerns and legal regulatory issues, were overcome through consultant studies and staff presentations which demonstrated positive cash flows, environmental benefits and minimal risks.

This project was initiated by the city when plans were underway for a new 1.0 million square feet convention center hotel complex. Planning began in 1982 with a study co-funded by PG&E and the California Energy Commission. The City of San Jose together with the Redevelopment Agency was fortunate in that from October 1984 to December 1986, funds from USHUD assisted in implementing the cost estimating and design phase. The project also required conceptual and technical understanding of the council before proceeding with approval for funding.

The major obstacles were complex coordination, funding, cost-effectiveness, and reliability concerns. A key factor in implementing this program was coordination of facility planning and cost estimating. This project had to track closely with all facets of design and construction of the facility in order to succeed.

At first, staff had to convince the City Council that the cost of providing the cogeneration system would be wise. In March 1987, the City Council approved the project. This decision was supported by a series of presentations to council demonstrating positive cash flows, minimal risks and rebutting counter arguments.

Negotiating with the hotel was the second major obstacle to be overcome. Legal hurdles involved crafting language acceptable to both the hotel and the city. Both parties wanted minimal risk and maximum benefit.

Cash flow and engineering quality control documents were drafted and revised until both parties reached agreement.

In summary, the project can be duplicated by other jurisdictions. It requires the attention of staff and consultants to complete feasibility studies, identify optimal sizes, propose funding and coordinate with the design, construction and operational phases.

# TECHNICAL DESCRIPTION
# OF THE COGENERATION SYSTEM

The cogeneration system installed in the San Jose Convention Center consists of a natural gas driven reciprocating engine manufactured by Cooper-Superior. The engine horsepower is rated at 2200, generating 1500 kW of electric power. The engine is a V-16 type, operating at 900 RPM. The in-house generated electric power and the PG&E supplied electric power is connected to one common main bus. From the main bus the power is distributed to six substations and the 100,000-sq.-ft. library building, as shown in Figure 29-3.

•   Substation A, supplies power to the electric chillers.

•   Substation B, supplies power to the motors.

•   Substations C, D, F, and G supply power to various portions of the convention center for lighting fans, escalators, air handlers, elevators, exhibitor requirements, and other office equipment.

•   The seventh main distribution panel circuit breaker supplies power to the library to operate the lighting, motors and other office equipment. The library does not run its boiler and chiller because the entire thermal energy is supplied by the convention center central plant and the cogeneration system.

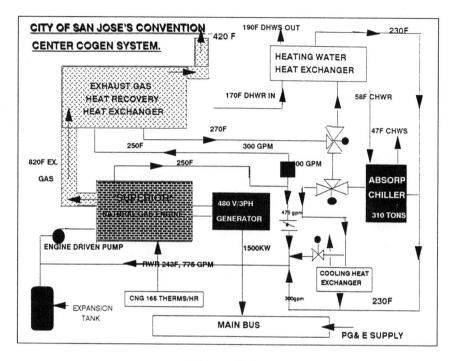

Figure 29-3.

The three-way valves, heat exchangers and the engine alarm of the cogeneration plant are directly integrated into the building's sophisticated "Rosemount" Energy Management Control System (EMCS). The EMCS also monitors the amount of energy supplied to the adjacent library building and the Hilton Tower.

# ENERGY BALANCE OF
# THE COGENERATION SYSTEM

The cogeneration system consumes natural gas at rate of 165 therms/ hr. The engine runs around the clock and around the year, except during the maintenance downtime. The electric power generated is about 1500 kW. The heat recovered through the jacket water cooling of the engine and from the exhaust gas heat exchanger amounts to about 6,000,000 Btus/hr, as shown in Figure 29-2.

# ELECTRIC POWER

The 1500 kW of electric power is fed directly through a step up transformer to the main bus, which also receives additional power from PG&E. From the main bus the power is distributed to several applications through six substations and the main circuit breaker. The approximate cost of producing the electrical energy is about $0.4/kWh.

# THERMAL ENERGY

Through advanced design techniques, heat is recovered from the engine jacket and the exhaust heat exchanger. The heat recovery rate is about 6,000,000 Btu/hr. The total heat recovered and used over one year is 5.1E10 Btus. At an average equipment efficiency of 85% the total Btus not purchased from the utility is 6.0E10. Monetarily, the heat recovered savings are about $150,000 a year.

**A Case Study of Energy Savings Comparison**
The City of San Jose's International Airport uses 7.10E+10 Btus of energy per year and pays $1.8 million to the utility. The convention center complex with its cogeneration system uses a total of 9.47E+10 Btus of energy per year and pays $768,000 towards its energy bills. Another way of analyzing is that the airport gets 39,282 Btus per dollar spent and the convention center gets 123,270 Btus for each utility dollar spent (about three times better). A comparison table is shown in Figure 29-4.
The City of San Jose's Water Pollution Control Plant has four cogeneration systems. These systems operate on a blend of digester and natural gas, thus saving the city hundreds of thousands of dollars every year.

**Natural Gas Purchasing Strategy**
In 1991, the California Public Utilities Commission approved the system for purchase gas in the open market. Since May of 1991, the city has purchased gas for cogeneration system from the open market reducing gas cost by approximately $8,000 per month. It is estimated that as the gas market becomes more competitive the savings will increase.

**Figure 29-4.**
**Energy Savings Comparison of San Jose**
**Convention Center and San Jose International Airport**
**(Fiscal Year 1990-91)**

| Bldg. | Energy Source | Elec. Energy (kWH) | Thermal Energy (Btus) | TTL Energy (Btus) | TTL Cost (Dollars) | Btus/ Dollar |
|-------|---------------|--------------------|-----------------------|-------------------|--------------------|--------------|
| SJIAP | PG&E | 1.64E+07 | 1.49E+10 | 7.10E+10 | 1.81E+06 | 39282 |
| SJCC | Cogen Plant | 1.28E+07 | 5.10E+10 | 9.47E+10 | 7.68E+05 | 123276 |

**Other Energy Conservation Measures**

In addition to the energy used by the convention center (produced by the cogeneration system), the city buys supplementary power from PG&E to meet the peak period demands. This is costing the convention center about $300,000 a year. However, the city is implementing several energy conservation measures to reduce the demand in buildings receiving energy from the cogeneration plant. Once these projects are installed, the peak period demand will reduce sizably, contributing to the savings further.

# Case Study: Cogeneration Operational Experiences— Quality Makes the Difference

*Chapter 30*

# Case Study: Cogeneration Operational Experiences— Quality Makes the Difference

*Paul L. Multari*
*Mission Operation and Maintenance, Incorporated*

T his chapter describes the experience of Mission Operation and Maintenance (Mission) in the operation and maintenance of combustion turbine cogeneration facilities. After a description of some of the facilities Mission operates, the capacity factor performance at each plant, and Mission's experience with turbine outages at these facilities are summarized. The chapter then focuses on the commitment to quality and discusses how quality operation and maintenance enhance the value of cogeneration assets.

## MISSION O&M SITES

### Kern River Cogeneration

The Kern River Cogeneration Facility (Kern River) was initially placed into commercial operation August 17, 1985. The plant is rated at 300 MW (ISO), utilizes four MS7001E combustion turbines which exhaust into four Struthers Wells heat recovery steam generators to produce 1.7 million lbs/hr of 80% quality steam. The steam is used for thermally enhanced oil recovery operations. These units can burn either natural gas or distillate fuel. $NO_x$ control to 42 ppm is achieved with water injection in the combustors. This plant is equipped with evaporative coolers on the inlet to improve performance during periods of high ambient temperatures.

### Sycamore Cogeneration

The Sycamore Cogeneration Facility (Sycamore) was placed into commercial operation January 1, 1988. This project, also rated at 300 MW (ISO), is configured with four MS7001E combustion turbines which exhaust into four Struthers Wells heat recovery steam generators. This facility also produces 1.7 million lbs/hr steam for enhanced oil recovery. Sycamore uses water injection in the combustors for emission control of $NO_x$ to a level of 42 ppm. Natural gas is the only fuel used at the Sycamore facility. The project is equipped with evaporative coolers.

### Harbor Cogeneration

The Harbor Cogeneration Plant (Harbor) consists of a single MS7001EA combustion turbine rated at 80 MW (ISO) which exhausts waste heat into a Combustion Engineering heat recovery steam generator designed to produce 465,000 lbs/hr of steam at 80% quality. The steam is used for enhanced oil recovery. The Harbor combustion turbine utilizes water injection to control $NO_x$ to a level of 42 ppm followed by selective catalytic reduction located in the boiler to further reduce the $NO_x$ emissions below 9 ppm. Harbor also utilizes a carbon monoxide catalyst to reduce CO emissions. The Harbor project began commercial operation on April 12, 1989, and can currently burn only natural gas fuel.

### Midway Sunset

The Midway Sunset Project (Midway Sunset) is rated at 225 MW (ISO). It consists of three MS7001E combustion turbines which exhaust into three Foster Wheeler heat recovery steam generators. The plant produces 1,000,000 lbs/hr of 80% quality steam used in enhanced oil recovery operations. The turbines are configured with General Electric's quiet combustor design which, with water injection, is capable of reducing $NO_x$ emissions to 25 ppm. The Midway Sunset project began commercial operation on May 8, 1989. These units can burn either natural gas or distillate fuels. The plant is also equipped with evaporative coolers.

### Midset

Midset Cogeneration Company (Midset) began commercial operation on May 1, 1989. The plant is configured with a single MS6001 combustion turbine, rated at 42 MW (ISO), which exhausts waste heat into a

Nooter/Eriksen heat recovery boiler and generates 215,000 lbs/hr of 80% quality steam. The steam is used for enhanced oil recovery. Midset also has an evaporative cooler and uses both water injection and selective catalytic reduction to reduce $NO_X$ emissions.

## Saguaro

Saguaro Power Company (Saguaro) reached commercial operation on November 22, 1992. This facility has two MS6001B combustion turbines rated at 38 MW exhausting into two Deltak HRSGs generating over 156,000 lbs/hr of steam. The plant is in combined cycle mode with a 25 MW GE steam turbine. Extraction steam (120,000 lbs/hr) is used for cogen at a nearby chemical plant. Saguaro has evaporative coolers, and massive steam injection plus selective catalytic reduction for $NO_X$ control.

## Sargent and Salinas

Sargent Canyon Cogeneration Company (Sargent) and Salinas River Cogeneration Company (Salinas) began commercial operation on February 22, 1992 and March 5, 1992, respectively. The plants each have a single MS6000 combustion turbine rated at 38 MW. Exhaust heat is recovered in a Nooter Eriksen HRSGs each producing 225,000 lbs/hr of 80% quality steam at 1500 psig. The steam is used for enhanced oil recovery. The plants are equipped with evaporative coolers and have General Electric's new "Dry Low $NO_X$" burners. $NO_X$ emissions are further reduced using selective catalytic reduction.

## Coalinga

Coalinga Cogeneration Company (Coalinga) went commercial on March 6, 1992. The plant has a single MS6000 combustion turbine rated at 42 MW, exhausting into a Nooter/Eriksen HRSG that produces 225,000 lbs/hr of 80% quality steam. The steam is used for enhanced oil recovery. The plant is equipped with evaporative coolers and uses water injection plus selective catalytic reduction for $NO_X$ control.

# MISSION O&M EXPERIENCE

The two tables below summarize some of Mission O&M's operating experience. Table 30-1 shows Capacity Factors (CF) for each plant Mission O&M operates. In the middle column of this table, Net Rated CF is shown and calculated as:

$$\frac{\text{Actual MWh Sold}}{\text{Rated Capacity} \times 8760 \text{ hours/year}}$$

This is the "Net Rated" CF for the machine, in contrast to "Net Contract" CF used for revenue purposes. "Contract" CF is calculated as follows:

$$\frac{\text{Actual MWh Sold}}{\text{Contract Capacity} \times 8760 \text{ hours/year}}$$

Contract Capacity is normally less than Rated Capacity and therefore "Net Contract" CF is normally higher than Net Rated CF and can even be greater than 100%.

The last column of Table 30-1 shows "On-Peak" CF for each location, which is calculated as follows:

$$\frac{\text{Actual On-Peak MWh Sold}}{\text{On-Peak Contract Capacity} \times \text{On-Peak hours/year}}$$

On-peak MWh sold is the most economically important measure of plant performance in all of our Power Sales Agreements, because power produced on-peak may sell for as much as ten times off-peak power

| | Table 30-1. Capacity Factors Since Commercial Operation Cogeneration Facilities Operated by Mission | | |
|---|---|---|---|
| Location | Commercial Start-up Date | Net Rated CF | Net On-peak CF |
| Kern River | 17-Aug-85 | 93.30% | 96.30% |
| Sycamore | 1-Jan-88 | 94.20% | 98.00% |
| Harbor | 12-Apr-89 | 93.00% | 98.30% |
| Midway Sunset | 8-May-89 | 93.50% | 99.90% |
| Midset | 1-May-89 | 91.70% | 102.80% |

price. Note that the On-Peak CF shown in Table 30-1 is based on contract capacity.

Table 30-2 shows a summary of turbine outages performed at Mission O&M operated plants since commercial operation. The numbers shown represent a significant body of experience.

Planned combustion turbine outages fall into three categories: Combustion Inspection (CI), Hot Gas Path Inspection (HGPI) and Major Turbine Overhaul (MTO). All of these outages are planned to occur such that the impact on revenue will be minimized.

**Table 30-2. Turbine Outages Performed**
**Cogeneration Facilities Operated by Mission**

| Location | Combustion Inspection | Hot Gas Path Inspection | Major Turbine Overhaul |
|---|---|---|---|
| Kern River | 28 | 7 | 4 |
| Sycamore | 16 | 6 | 2 |
| Harbor | 5 | 1 | — |
| Midway Sunset | 7 | 2 | — |
| Midset | 3 | — | 1 |
| Totals | 59 | 16 | 7 |

As a general rule, these outages are planned in the non-peak months of the year. Short duration outages, such as the CI, can be accomplished in 2-3 days and are often scheduled for weekends. The HGPI typically takes from 5-10 days and MTO takes 4-6 weeks.

The particular features of the Power Sales Agreement at each plant will determine the strategy for each outage. For example, at most of our plants, it is most profitable to minimize down time by working weekends and using 24-hour coverage. There are certain projects, however, where it is not economic to pay premiums to work around the clock and on weekends or holidays to accomplish an outage. At the time of this writing, we are conducting a CI at Midset that is planned for five days, working only day shift, Monday through Friday, because the marginal revenues from accelerating the outage would not offset the marginal costs.

Some Power Sales Agreements provide that certain holidays occurring during the on-peak season are not on-peak days. We have, on occasion, used these holidays to accomplish maintenance during the peak season.

As we have become more accomplished at performing turbine outages, the time required for the turbine work has decreased. We have conducted combustion inspections on Frame 7s in as little as 23 hours. However, as the plants age, auxiliary equipment, instrumentation, and boilers required more attention. Thus, the turbine work on a combustion inspection does not determine the timing of the outage. For example, a weekend outage will use the full 48 hours for miscellaneous maintenance, even though the turbine work can be done in 30 hours or less.

# MISSION O&M'S COMMITMENT TO QUALITY

Mission Operation and Maintenance is committed to provide the highest quality operation and maintenance service in the independent power industry. To do so, we must provide safe, reliable, and cost effective service to our customers.

In maintenance, this means understanding the cost of "downtime" to the project, and minimizing those costs for a desired level of plant availability or capacity.

# MINIMIZING DOWNTIME COSTS

Downtime costs have two components: direct costs and indirect costs. Direct costs are labor and materials spent on maintenance tasks. These are the costs that appear in a plant's maintenance budget. The indirect component contains the lost energy and capacity revenues. These costs are incurred by the company, but are not usually part of the plant's maintenance budget. Thus, recorded maintenance costs give an incomplete picture of total costs. An analysis of the cost to achieve a certain level of availability must include both direct and indirect costs.

This is sometimes referred to the "cost of poor quality." The losses that a company suffers from poor O&M may not always be evident in its budget. The value of high quality O&M service is in reducing these hidden costs. Analysis of the cost of planned and unplanned downtime starts with a thorough understanding of the plant's Power Sales Agreement.

Planned downtime is less costly than unplanned downtime because the timing of planned downtime and associated indirect costs, can be controlled. Planned downtime is typically scheduled during off-peak hours when revenue losses are minimized.

Figure 30-1 illustrates the relationship between planned maintenance and the cost of planned downtime. The figure shows both indirect and direct downtime costs and these costs are assumed to increase linearly as each increment of maintenance is added. This means that planned downtime occurs when incremental revenues just equal variable costs.[1]

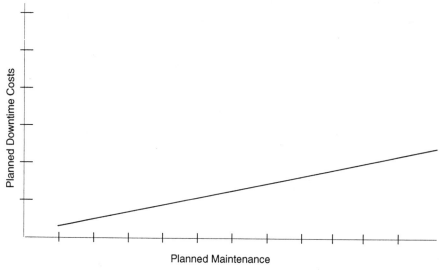

Figure 30-1. Planned Downtime Costs

---

[1] Not all planned maintenance requires downtime. Much preventive and predictive maintenance that avoids unplanned downtime can be done while the plant is on line. This simple analysis addresses only the planned maintenance that requires downtime.

Unplanned downtime is usually caused by breakdowns, and has higher direct costs because: 1) many times the equipment is damaged and requires additional labor and material inputs to repair; 2) an emergency exists and crews are working overtime and parts may have to be expedited. More importantly, indirect costs are incurred due to lost energy and capacity revenues. Clearly, if unplanned downtime occurs during off-peak hours, indirect costs are low. For the purpose of this paper, unplanned downtime is assumed to occur randomly at all periods of the year and the indirect cost of unplanned downtime includes a "weighted average" cost of lost revenue for energy and capacity.

Figure 30-2 illustrates the relationship between planned maintenance and unplanned downtime costs. These costs include both direct and indirect components. The graph shows that increased planned maintenance initially decreases unplanned downtime costs very rapidly. This is because maintenance can reduce the rate of non-random failures. At some point, however, further planned maintenance cannot significantly decrease unplanned downtime costs, as shown in the figure.

The total costs of any combination of planned and unplanned downtimes is the vertical sum of Figures 30-1 and 30-2. This is total direct and indirect costs and is graphed in Figure 30-3. As planned maintenance increases from zero to Point X, unplanned downtime costs

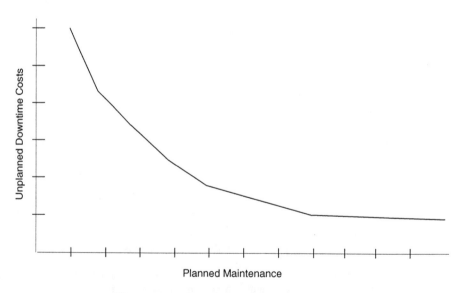

**Figure 30-2. Unplanned Downtime Costs**

drop from Point A to Point B, while planned downtime costs increase from Point C to Point D. Beyond Point X, further planned maintenance is excessive and its costs will not be offset by reduced unplanned downtime costs. Note that totals are minimized at Point E.

## CONCLUSION: ENHANCED ASSET VALUE

The operating results that have been achieved at plants Mission O&M operates are due, in part, to successful management of outages. High Quality O&M services begin with a thorough understanding of the Power Sales Agreement. Analysis of direct and indirect costs of downtime minimizes total maintenance costs while achieving desired levels of output. This enhances the value of the cogeneration asset to its owners.

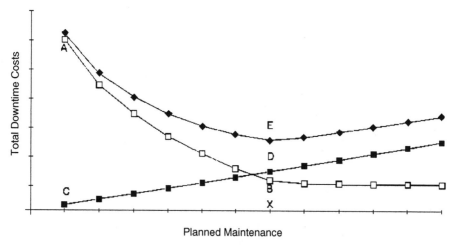

**Figure 30-3. Total Costs**

*Chapter 31*

# Case Study: Ocean State Power, The First "Independent Power Project," Sets an Example in Social Responsibilities

## Chapter 31

# Ocean State Power, the First "Independent Power Project," Sets an Example in Social Responsibilities

*Jackie D. Moran*
*Ocean State Power*

---

**J**ust because a combined cycle cogeneration/IPP project makes economic, environmental, technical and regulatory sense doesn't mean that it will arrive quickly, without hassle, into full production. More and more, the role of the local communities must be understood, and these communities must be made part of the planning process. Unless this is accepted, they may become understandably recalcitrant, as evidenced by the spreading "NIMBY" (Not In My Backyard) syndrome.

Ocean State Power* wisely understood the correct way to engage local communities in the planning state and has thereby benefited, as this article reports.

---

*OSP is a general partnership comprised of these companies.

**J. Makowski Company,** a Boston-based energy project development and management firm with approximately 20 years of experience in the gas and electric industries.

**TransCanada PipeLines, Ltd.,** a holding company and one of North America's largest gas transportation and marketing companies, based in Calgary, Alberta.

**Eastern Utilities Associates (EUA),** a Boston-based holding company whose affiliates are principally engaged in the generation, transmission, distribution and sale of electricity to approximately 290,000 customers covering a 595-square-mile area in Massachusetts and Rhode Island.

**New England Electric,** a holding company headquartered in Westborough, Massachusetts. Among its subsidiaries are three retail operating companies which together serve over 1.2 million customers in Rhode Island, New Hampshire, and Massachusetts.

431

The OSP Project received a Special Recognition Award at the Cogeneration Congress, Boston, June 12, 1991, and was one of six plants in the nation honored by Power Magazine with the 1992 Powerplant Award for applying advanced combined-cycle technology to a dispatchable independent powerplant and for demonstrating state-of-the-art zero-discharge water treatment.

# PROJECT OVERVIEW

Ocean State Power (OSP) is a natural gas-fired, 500 Megawatt (MW), combined cycle electric generation facility in Burrillville, Rhode Island. Unit I of Ocean State Power began commercial operation in December, 1990 and Unit II in October, 1991. The four partners in the $430 million, twin-unit project negotiated firm 20-year power sale contracts with four investor-owned utilities in Massachusetts and Rhode Island.

The project, which has been hailed by the Federal Energy Regulatory Commission (FERC) as a model for the emerging Independent Power Producer (IPP) industry, represents a number of milestones and innovations for the electric and gas industries:

*   OSP is the first baseload power plant to be constructed in New England in more than a decade.

*   OSP is the first power plant built in New England to exclusively use Canadian natural gas.

*   OSP is one of the first major generating facilities to contract for a firm, 20-year gas supply directly from producers and to arrange for firm transportation to the plant.

*   OSP's fuel price is tied to the average fossil fuel cost for the New England Power Pool, thus facilitating its dispatch as a baseload unit.

# BECOMING A "GOOD NEIGHBOR"

As part of their corporate responsibilities, Ocean State Power staff members have become deeply committed to the concerns of their neighbors, the communities of Burrillville, RI and Uxbridge, MA.

From the onset of the project, Ocean State Power has been committed to being a Good Neighbor and a contributing member of the community. It is seen as a corporate responsibility and is inherent in the philosophy of the company.

In the early stages of the project, a public relations consulting firm was contracted, Duffy and Shanley of Providence, to ensure that the low-key, informational and personal contact approach desired by the partnership was effectively implemented. Thus the groundwork and credibility base was established in conjunction with key Makowski personnel.

The J. Makowski Company, project developer, continues as managing partner for the project. The community and public relations manager joined the project as construction got underway, building on the established base and introducing new community involvement programs that augmented the pre-existing company philosophy.

Adhering to this policy and philosophy, the power plant, by design, has a low visibility factor in the surrounding area Often first-time visitors to the facility drive past the site and end up at a pay phone in town asking: "Just where is the plant?" Equally, by design, OSP staff members are highly visible in town, engaged in community activities. Here are some examples:

OSP staff members are in the classrooms with an Adopt-A-School program, a business/education partnership under the auspices of the local Chamber of Commerce, discussing energy, the environment, electricity and engineering with school children from the William L. Callahan School, grades 3 through 5. Sometimes special sessions are held for teachers and guidance counselors.

In 1991, OSP reached out to various contacts and resources and provided 12 speakers, most from other countries, for an International Reading Week program at a local elementary school. The program has become an annual event. Another ongoing program is mentoring which pairs an OSP employee on a regular weekly basis with a pupil in need of extra attention.

OSP members present awards at the annual town arts festival, attend dedications of local ballfields, provided support for local parades and are involved in additional area activities, as they arise. The plant annually sponsors a team of employee riders in a two-day, three-state, 150-mile bike tour to benefit the Rhode Island Chapter of the Multiple Sclerosis Society.

OSP also sponsored the 1990 RI Public Television's Family Fun

Photo credit—Sky High Enterprises/RI

Ocean State Power Units I and II
The first year cost of power, calculated using traditional utility ratemaking principles, was approximately 6 cents/kWh given current fuel prices. Electricity from OSP was among the most economical within the New England Power Market.

Day, a fund-raising event designed to keep public television available to Rhode Island viewers. OSP personnel attend banquets commemorating anniversaries of local, volunteer fire-fighting units; participate in Cub Scout-sponsored Safety and Health Fairs, mix and mingle with townspeople at the library open house. They are present at local graduations and award ceremonies and regularly are involved in classroom presentations, programs and site visitations for high school, vocational school and college students.

An annual event since early construction days has been the Holiday Food Drive for the local needy. Last year 19 gift-wrapped boxes of food, along with a monetary donation from individual employees, was do-nated to a food kitchen run by the area ecumenical council of churches. For the past two years, the holiday project was shared with OSP's business/education partner, the W.L. Callahan School.

# THE "SNOWMAN" PROJECT

A new "Snowman" project for needy children was begun in 1991 with the plant Christmas tree decorated with six-inch plain white snowman silhouettes, created by the art teacher at a local school. Each snowman had the name and age of a child, his/her "wish list," and the ages of siblings. Employees selected a snowman, personally purchased and wrapped appropriate gifts and when the gifts were brought to the plant, the snowman returned to the tree, this time with a bright red heart.

Not wanting to unduly burden employees who had already made the earlier food drive a success through their generous donations, management initially placed only 12 snowmen on the tree. An hour later, there were requests for more. A total of 38 children and their siblings benefited.

Next year, there will be more. There are 98 employees at the plant; some gave in groups, others in departments and still others as individuals. All gifts were anonymous, "With love from Santa" and town employees volunteered their time to distribute the many festively wrapped items. The warmth that bathed the site and its workers ensured that Project Snowman was the start of a new OSP holiday tradition.

# MAKING FRIENDS WITH PROPERTY OWNERS AND TOWN OFFICIALS

When Ocean State Power was first proposed, it was given a positive reception by town officials, which doesn't mean they had no questions. Many meetings were held with groups and individuals to explain specifically what the project was—and what it was not.

As a new business in town, OSP was able to bring significant economic benefit to the community without adversely affecting the quality of life or health or safety of the citizens of Burrillville. Certainly, this had a positive impact. A negotiated tax treaty will provide $73 million to the town over a 20-year period.

The company has always taken a pro-active, upfront and forthright stance in dealing with the public. Perhaps most importantly, OSP people listened, and continue to do so… to comments, to questions, to concerns. Then, they follow through with answers and actions designed to address what was heard.

To that end, regular meetings, about every six weeks during construction, were held at the site with neighbors in the area surrounding the plant. Since commercial operation, meetings are scheduled about every 10-12 weeks.

Invited are about 65 nearby property owners. Attendance seldom exceeds 20 neighbors, but they have told us they like to know the option is available. The no-holds-barred discussions are interesting and informative to all parties, but very much a friendly and neighborly get-together. The meeting finishes with an "up close and personal" tour of the facility.

An average of 100 visitors tour the power plant each month, some in groups of two or three. Among them are neighbors and townspeople. Periodically, tours are scheduled for specific segments of the town populace, such as town officials, local firefighters, educators, library trustees and various clubs and organizations. In addition, OSP speakers travel off-site to address community groups.

The entire plant workforce (98 people, including management) is mindful of Ocean State Power's community commitment. To carry out its comprehensive community relations program requires the efforts of many. Recently, a Recognition Breakfast was held to honor 33 employees for outstanding assistance and contributions to the community relations effort. Cooperation is given at all levels of management, operations and maintenance and it is critical to a meaningful and successful program.

# THE IMPORTANCE OF LOCAL EMPLOYEES

A big question posed by town officials was how OSP would view townspeople when filling positions at the plant. To address that concern, local people were interviewed, and hired, for a number of key positions, including the public and community relations person and administrative staff.

When it was time to hire an operations and maintenance workforce for Unit I, about 55 jobs, an informational night was held at a local hall. Almost 1000 people attended; the lines stretched out into the street. Job description blow-ups were posted throughout the hall, an ongoing slide show presented a documentary of the project, and everyone got to speak to an OSP representative. It was not until a week later that employment ads ran for public notification.

Most of the positions required highly technical expertise. When all was done, with well over 1000 applications received, 24% of the workers hired were local; that figure is now 34%. An interesting story involves one local individual who was turned down for Unit I and told that additional training might qualify him the following year for Unit II. Andy, on his own initiative, with suggestions from the OSP personnel manager, took the required courses, reapplied and is now a much valued operator, hired when Unit II began its employee search.

# SCHOLARSHIP AND COMMUNITY FOUNDATION

Among the tangible benefits to the community are the Scholarship and Community Foundations established by Ocean State Power. Based on the projected lifespan of the plant, for 20 years a total of $100,000 is given annually to the host community of Burrillville, RI and to the abutting Town of Uxbridge, MA:

- $50,000 in outright scholarships for resident graduating seniors.

- $20,000 in a scholarship endowment fund which will activate in 20 years.

- $30,000 in grants for community and civic projects.

The two foundations are administered by separate committees of townspeople, each with only one OSP representative. The Community Foundation has a wide and diverse impact. Written proposals are solicited from townspeople and acted upon by the committee.

Examples of the types of projects which have been funded include:

-Community walkway along historic canal route
-Training classes for visiting nurses
-Police radio and radar units
-Library renovation and expansion
-Seasonal display lighting for the common and downtown areas
-Adult literacy programs
-Regional rape crisis services
-Trees for the local conservation program
-Lighting of community athletic fields
-Handicap access in the Veterans building
-Playground equipment
-CPR talking mannequin for training emergency medical technicians
-"Jaws-of-life" for local rescue unit
-Pagers for emergency medical technicians
-Expansion of the local animal shelter

Landscaping for the plant site focused on keeping with the ambience of the host environment. A 300-foot buffer of trees separates the plant and abutting properties. On the site, a small historic cemetery and a section known as Crow's Hollow (thought to be archaeologically significant) now receive once-needed maintenance attention.

During plant construction, special measures were taken to fence off the areas. With the project completed, markers will be erected designating the areas, in coordination with the local Historic Society. Implementing of the latter two projects is ongoing.

The story of OSP is exciting and dynamic; it is also one of caring and commitment to the spirit of community.

*In return, the two communities of Burrillville and Uxbridge have welcomed the twin-unit facility as important supporters of their social and economic fabrics. OSP's intelligent, sensitive partnership efforts are a model for any major industrial expansion.*

# TWO IMPORTANT
# OCEAN STATE FOUNDATIONS

## Ocean State Power Community Foundations

Twin foundations in Burrillville, RI, and Uxbridge, MA, are separate, private, nonprofit corporations funded by Ocean State Power for the purpose of aiding and contributing to civic projects in the respective towns.

Established in 1988, the first grant was in the amount of $15,000 to each community. This amount was increased to $25,000 in the second year. In 1990, the year of commercial operation for Unit I of the 500 MW plant, the program was fully implemented with contributions of $30,000 to each of the two towns.

Contributions in this amount of $30,000 will be given over the 20-year life of the plant. The program will terminate in the year 2010, at which time more than $1 million will have been awarded for the betterment of the local communities through civic project grants.

In Burrillville, a nine-member committee administers the distribution of funds for the program. The Uxbridge committee is comprised of five members. There is one Ocean State Power representative on each of the foundation boards.

### Ocean State Power Scholarship Foundations

Burrillville, RI, the host town for the Ocean State Powerplant, and Uxbridge, MA, the plant's abutting neighbor, are each included in the Ocean State Power scholarship programs. Twin foundations function separately as private, non-profit corporations funded annually by OSP for the purpose of providing scholarships for deserving high school graduates whose permanent address is in one of the two towns.

Established in 1988, the first grant was in the amount of $35,000 to each community. The contribution was increased to $50,000 in 1989 and will remain so for the next 20 years, with the addition of $20,000, to each town, beginning in the year of commercial operation (1990). The additional monies are for the scholarship endowment fund which will be activated after the 20-year period, which is based on the 20-year life of the plant.

Funding will cease in the year 2010, at which time almost $3 million will have been contributed by Ocean State Power to qualified students pursuing higher education.

Company officials have deemed the program "an investment in the future of the community. "

In Burrillville, a nine-member committee administers fund distribution. The Uxbridge committee is comprised of six members. One Ocean State Power representative sits on each of the foundation boards.

# Index

441